→INTRODUCING

TIME

CRAIG CALLENDER & RALPH EDNEY

Published in the UK in 2010
by Icon Books Ltd.,
Omnibus Business Centre,
39-41 North Road, London N7 9DP
email: info@iconbooks.co.uk
www.introducingbooks.com

Sold in the UK, Europe, South Africa
and Asia by Faber and Faber Ltd.,
Bloomsbury House,
74-77 Great Russell Street,
London WC1B 3DA
or their agents

Distributed in the UK, Europe,
South Africa and Asia by TBS Ltd.,
TBS Distribution Centre,
Colchester Road, Frating Green,
Colchester CO7 7DW

This edition published in Australia
in 2010 by Allen & Unwin Pty. Ltd.,
PO Box 8500, 83 Alexander Street,
Crows Nest, NSW 2065

Previously published in 2001

This edition published in the USA
in 2010 by Totem Books
Inquiries to: Icon Books Ltd.,
Omnibus Business Centre,
39-41 North Road,
London N7 9DP, UK

Distributed to the trade in the USA
by National Book Network Inc.,
4501 Forbes Boulevard, Suite 200,
Lanham, Maryland 20706

Distributed in Canada by
Penguin Books Canada,
90 Eglinton Avenue East, Suite 700,
Toronto, Ontario M4P 2Y3

ISBN: 978-184831-120-6

Text copyright © 2001 Craig Callender
Illustrations copyright © 2001 Ralph Edney

The author and artist have asserted their moral rights.

Originating editor: Richard Appignanesi

Printed by Gutenberg Press, Malta

What is Time?

The great theologian and philosopher, **St Augustine** (AD 354–430), famously wrote of his puzzlement in *The Confessions*.

After pointing out all the things he is able to say about time without knowing what it is – for instance, that it takes *time* to say this – he admits that he really is in a "sorry state, for I do not even know what I do not know!".

Augustine is not alone in his bewilderment. The question of what time is and related puzzles – such as whether the past and future are real, whether time travel is possible, and the explanation of the direction of time – are among the most intractable yet fascinating questions asked.

All Kinds of Clocks

In everyday life, we are probably most familiar with time from two sources: clocks, and our inner psychological experience of time.

Clocks are everywhere. There are grandfather clocks, watches, alarm clocks, even incense clocks that let you tell the time through scent.

There are also natural clocks.

But clocks existed well before the modern invention of portable artificial ones.

Over four thousand years ago, the Egyptians used obelisk shadow clocks, sundials, and water clocks which measured time by the flow of water passing through a stone vessel.

By 1800 BC, the ancient Babylonians had divided the day into hours, the hour into sixty minutes, and the minute into sixty seconds.

All the great civilizations of the past used the positions of the sun or stars to tell the time.

These clocks were very accurate.

Looking at the stars with the naked eye, an ancient astronomer could tell the time to within fifteen minutes. And anyone can tell roughly the time merely by looking up at the sun.

Biological Clocks

We also carry within us our own biological clocks. The human heart beats 70 times on average each minute. Our moods, alertness, and appetites follow regular patterns, depending on the time of day, the lunar cycle or the season.

Our biological clock seems to be intimately associated with a group of nerve cells in our brain's hypothalamus.

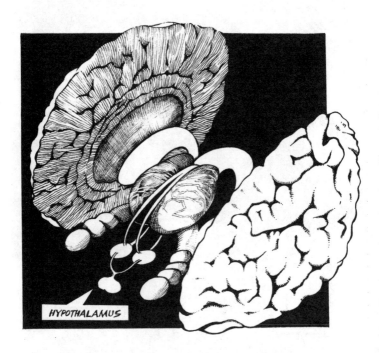

HYPOTHALAMUS

These cells are linked to the retina of our eyes and appear to regulate cycles of hormone secretion, our skin temperature and cycles of rest and wakefulness. The hormone melatonin is thought to play a significant role in controlling our daily (circadian) rhythm.

Biological clocks are not special to us. Every creature in nature seems to have them. Some are so good that they have been proposed for human use. The Swedish naturalist **Carl Linnaeus** (1707–78) thought we might use flowers as clocks.

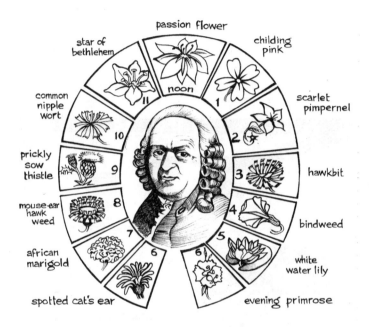

Surprisingly, not every biological clock is based on the day, lunar cycle, season or year. The cicada is a remarkable chirping insect that remains underground for 17 years. Then 17 years after going underground, thousands of them emerge at the same time, climb the trees, mate and then die a few hours later, whereupon the 17-year cycle begins again.

Whether natural or artificial, clocks have helped order the patterns of life for as long as human beings have been around. In modern society, they also can cause a lot of stress.

Psychological Time

We also *feel* time pass. In addition to the physical time measured by various clocks, there is also psychological time. We have memories of the past and anticipations of the future. And we experience temporal durations of different sizes. We are personally, subjectively aware of time passing.

Everyone can guess roughly how much time has passed between two events.

Some people can do this surprisingly well, as if there were little inner clocks in our heads – related somehow to the biological clocks.

The interesting thing about these inner clocks is that they seem to speed up or slow down for a person in ways that disagree with other people's inner clocks.

According to a watch, the trip on a super-fast roller coaster might take only 11 seconds.

11 seconds might seem an eternity to the person on the ride, whereas it may seem like almost nothing to someone waiting. A game of basketball might seem to pass in no time to the child playing it, but forever for the parent watching his twentieth such game in a month!

To begin our investigation into time, it's important to see that time is more than merely clocks or the subjective experience of time. Time isn't simply the alarm clock on your nightstand or something solely in your mind. Once we establish this, curious and deep questions will be right around the corner.

Is Time Merely in the Head?

After calming from his initial panic, Augustine argues that time doesn't really exist outside the head.

The Persian philosopher **Avicenna** (980–1037) agreed with him.

And **Henri Bergson** (1859–1941), the French philosopher, argued for this position too.

Can this be right? Although people disagree about their feelings of how much time has passed, they also enjoy remarkable agreement about the temporal ordering of events.

For example, the father and son returning home from the basketball game might not have looked at a clock since they left for the game – and maybe it was overcast, so they have no sense of where the sun is.

Suppose they guess what time it is before actually looking at a clock. Their guesses might disagree by as much as a couple of hours. They might even argue about who is right, but they typically won't argue much about the *ordering* of events that took place.

"We agree that the free throws by Smith in the second half occurred sometime after his free throws in the first half . . . "

"And Joey broke his finger when Smith stepped on it."

Except in rare circumstances, everyone (who has the same information available) agrees – for the most part – on the *time order* of events. There is definitely something objective and independent of a particular person's feelings about the time ordering. The objectivity of the ordering of events in time proves that there is more to time than just our psychological sense of its passage. There is the fact that events seem to be laid out in a unique and observer-independent succession in time.

Clocks and Time

Is this agreement merely agreement about what clocks will say? Maybe all there is to time *is* clocks. This is actually already a deep question. But, at least at first glance, it seems the answer is "no", for we often talk about a clock being *wrong*. You might say my watch is ten minutes slow or even completely off. This may be your excuse for being late for an appointment. But is your watch an infallible guide to time? No, we know it will "lose" a few seconds per year, even if it's pretty good.

Between each "tick" of the clock, we want the same amount of time to pass. It should be no surprise that pendulums, which have regular periodic motion, can be used as clocks. But pendulums aren't perfect. On a boat in high seas their motion will be disrupted, or in hot weather they may behave differently than in cold weather.

Consider a pendulum swinging back and forth twice. How do we know that the amount of time that passed on its first trip back and forth is the same as the amount of time that passed on its second trip? This question illustrates what the German philosopher **Hans Reichenbach** (1891–1953) called the "problem of the uniformity of time".

YOU MIGHT ANSWER THAT THE FIRST TRIP SIMPLY *FEELS* AS LONG AS THE SECOND

BUT THIS ANSWER ISN'T VERY SATISFACTORY FOR THREE REASONS

Firstly, your personal estimations of time won't be precise enough for science. We need to know whether the first trip seemed *exactly* the same as the second trip. Secondly, your feeling as to the amount of time that passed is subjective. You might say the same amount of time went by, but your friend might not think so. Thirdly, and most importantly, you're measuring the time that passed with your thoughts, but these are – plausibly – physical processes, and so this merely pushes our question back a step. That is, we would then ask how you know how long your thoughts *last*?

How Long is an Interval of Time?

We can't directly measure durations of time passing. *We never measure pure time.* Is this minute the same length of time as the next minute? In one sense the answer is of course yes: minutes are defined to be the same length of time. But we mean something deeper.

Back to the swings of the pendulum. Despite our inability to directly measure time-lapse, we still think a pendulum can be wrong. Why? Well, suppose some troublemaker wanted to treat his pendulum as an infallible guide to time. What would be wrong with this?

Imagine that he decided to take it to the equator by boat. Even discounting the rocking of the boat, we can expect at least two other factors to affect the pendulum: air at the equator is more humid and provides greater resistance to the pendulum, and the gravitational field that attracts the pendulum is

slightly weaker at the equator. By our standards, the pendulum *slows down*.

He is going to have to say his boat is moving faster than before, even though (let's assume) it has the same-strength wind on its sails, same-strength current, etc. He must explain why all the clocks in the world are magically starting to speed up; why the speed of the sun has altered. Since he can't provide an explanation for these changes, and yet we can, it seems we're right and he's wrong. Our hypothesis, that time can be given by the motion of the stars, for example, is a scientifically better claim than his pendulum hypothesis.

The Most Reliable Clocks

Let's first pause to note that three clocks have proved to be very good. Historically the sun and the night sky have been most important.

The sun defines regular "ticks" of a clock, if we think of it as ticking each time it crosses the meridian. The stars in the night sky define "ticks" through a chosen star's passing through due south. Both clocks are better than my wristwatch.

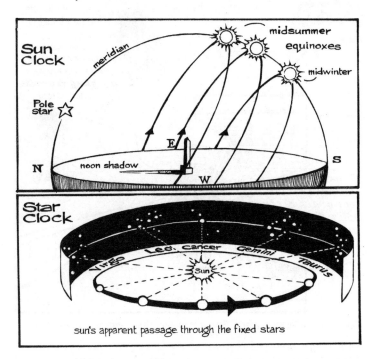

sun's apparent passage through the fixed stars

I can explain a discrepancy between the two by appealing to low batteries in my watch, whereas I cannot blame the discrepancy on the sun or stars slowing down or speeding up. Though these two clocks are amazingly accurate (the night sky being better than the sun), there are better still.

The Atomic Clock

The development of particle physics in the 20th century, and in particular ideas by the American physicist **Isidor Rabi** (1898–1988), gave us the atomic clock in 1949. All atoms have a so-called natural resonance frequency, and this extremely regular oscillation can be used to define the "ticks" of a clock. Atomic clocks have proved more regular than solar or astronomical clocks. In 1999, the National Institute of Standards and Technology in Boulder, Colorado (US) started using an atomic clock known as the NIST F-1 to define the second.

A second is defined precisely to be 9,192,631,770 vibrations of the cesium atom! The NIST F-1 (with a similar clock in Paris) is the most accurate clock in the world today. With a pool of atomic clocks around the globe, it is used to help define Coordinated Universal Time, which in turn helps define the speed of light, the length of the standard metre, and so on. But even this incredibly accurate clock will "lose" nearly a full second every 20 million years. Nothing is perfect!

Absolute, True and Mathematical Time

If the NIST F-1 is the best physical clock that we know in the universe, how do scientists know it will lose time, even only 1 second every 20 million years?

The answer is that the laws of physics tell them so. This is a deep point, made best by none other than **Sir Isaac Newton** (1642–1727) – the father of classical mechanics and perhaps the greatest physicist of all time. Time, Newton says, should not be confused with its sensible measure. By "sensible measure", Newton means the actual clocks we happen to use.

Real time, according to Newton, does not depend on any particular clock, or even any particular material object in the universe. Time is independent of the contents of the universe.

It is this time that is used in the unchanging laws of physics. The laws of physics tell things *where* to be and *when* to be there. In telling them when to be where, Nature assumes a particular time measure.

For example, classical physics says that the acceleration of a freely falling body is constant.

19

True Time

Implicit in the very law of nature is an ideal perfect clock. The true time is the one that makes this law true. If we adopted some non-standard measure of time, then this law would not be true. If we threw a rock from a rooftop and measured time with our explorer's pendulum, we would get different results for the same experiment in England and on the equator.

That is how scientists know that even the near-perfect F-1 clock will lose time. The current laws of physics tell them so. And while there will be better and worse physical examples of clocks, we can't expect any of them to perfectly match the ideal perfect clock according to which the universe is governed.

According to Newton, we shouldn't confuse any of these actual imperfect clocks with the perfect, invisible, clock that is independent of any physical object: Time.

Not everyone agrees with Newton. His idea of absolute time continues to be both influential and hugely controversial. In the language of philosophers of science, Newton is both a *Realist* – he thinks that the time mentioned in the laws of physics is really Time itself – and an *Absolutist* – he thinks that time is independent of any particular physical process.

Opponents of Newtonian Time: Relationalism

Opponents of Absolutism, known as Relationalists, hold that time is essentially just *change*, or the measure of change. By change we mean change in the relationships between physical objects. **Aristotle** (384–322 BC), the Greek philosopher, held that time is simply the measure of motion. Time is the measure of one physical process against another.

In this view, contrary to Newton's, time is dependent on the physical contents of the universe since time is defined via their change. Time for Aristotle *is* dependent on its sensible measure – actual physical clocks.

In the Relationalist view, because time is dependent on *physical movement*, it seems time doesn't pass when there is no change. Can we conceive of looking at the stars, having them stop, and still being able to experience time passing? Aristotle considered this question and pointed out that in such a case we're still measuring the progression of time with our changing thoughts and feelings. We need these to stop too.

Because our brains will be frozen too, it is true that we *wouldn't notice* the passing of time. Could time pass by nonetheless? It would, if time is independent of change. So, in Newton's view, it would at least be conceivable for time to pass without any change at all. But according to Relationalism, this is impossible. Time is just the measure of change. No change, no time.

A Scenario of Time without Change

It is possible at least to conceive of having good reason to say time passes without any change. The American philosopher **Sydney Shoemaker** invented a scenario in which it seems we have empirical reason to say that time passes without change.

Consider a world made up of three regions – three galaxies, say – A, B, and C. Imagine that the inhabitants of each region can observe and communicate with each other.

They notice that these local freezes occur at regular intervals: galaxy A freezes every 3rd year, B every 4th year, and C every 5th year. Given these rates of freezing, one would expect that C should observe a combined A and B freeze every 12th year. This is what happens, and C tells A and B about it.

Simple arithmetic then dictates that A, B, and C should together freeze every 60th year. But since they make up the entire world, that means there is reason in such a world to believe that every 60th year the entire world freezes; that is, that time passes for a year without a single change.

Though this example hardly proves relationalism false, it does show that it's possible to have reason to believe time passes without change.

Can Relationalism replace Absolute Time?

Gottfried Wilhelm Leibniz (1646–1716), a German mathematician and philosopher, independently invented the branch of mathematics known as calculus and fought viciously with Newton over who should get proper credit. In a famous debate, Leibniz also objected to Newton's view of time, arguing that if Newton were correct then it would make sense for the entire universe to have come into being one second earlier or later than it did. Leibniz didn't worry so much about the untestability of this hypothesis.

Without sufficient reason, Leibniz thought, God wouldn't have created the universe at all! Minus the theology, the point is really that the claim of absolute time leads to unnecessary possibilities.

THEOLOGICAL MATTERS ASIDE, I ARGUE THAT MY TREMENDOUSLY SUCCESSFUL PHYSICAL THEORY **REQUIRES ABSOLUTE SPACE** AND **TIME**

RELATIONALISTS MIGHT NOT LIKE THIS...

...BUT UNTIL AN ALTERNATIVE RELATIONAL THEORY IS SHOWN TO WORK, **ABSOLUTE TIME** IS THE BEST

Newton was keen to stress that his physics illustrated the greater glory of God. He didn't assume God would be so stymied by a choice between two identical situations that He couldn't do anything at all – like Buridan's famous example of an ass who starves while trying to decide between two identical bales of hay.

Since the Newton–Leibniz debate, many relationalists have mounted other arguments against Newton's absolutism. In the 20th century, philosophers and physicists have squabbled over whether Einstein's general relativity is relational or absolutist. The debate thus continues within the context of a new physical theory.

Conventionalism

Newton's "realist" conception of time also draws opposition – his idea that there really is one true time "out there". A prominent opposing school of thought is known as Conventionalism. Reichenbach, mentioned earlier, was a prominent conventionalist. **Henri Poincaré** (1854–1912), a French physicist, philosopher and foremost a mathematician, stated the conventionalist's view of time very succinctly. After considering the way astronomers measure time he summed up their approach with approval . . .

TIME SHOULD BE SO DEFINED THAT THE EQUATIONS OF MECHANICS MAY BE AS SIMPLE AS POSSIBLE

THERE ISN'T ONE WAY OF MEASURING TIME MORE TRUE THAN ANOTHER: THAT WHICH IS GENERALLY ADOPTED IS ONLY THE MORE CONVENIENT

Poincaré doesn't think that there is one true time "out there". But he isn't thereby committed to believing that any clock is as good as another. The one used by the simplest physical theory is the best one.

If we used a pendulum on a boat as our arbiter of time, then it would be very hard if not impossible to have a science of the world. Even if we managed to scrape one together, still it would be a complicated mess. So some measures of time are more convenient than others.

If some other physicist, Snewton, say, invented Snewton's physics, and this new physics was simpler and had a different measure of time, then that would be what we should refer to as time. There is no fact of the matter regarding which time is the true time. Simplicity, like beauty, is in the eye of the beholder. Which measure of time we use is a matter of convention, not fact, according to this view. For some, this consequence alone is enough reason to reject Conventionalism.

A Universe Out of Sync?

In the 1930s, the physicists **Paul Dirac** (1902–84) and **Arthur Milne** (1896–1950) independently shared a worry that is interesting to mention at this point. The worry was that we might have to introduce more than one time scale.

The atomic clock is better than the pendulum as a timekeeper – we can explain discrepancies between the two by saying the pendulum is at fault due to changing frictional and gravitational forces.

BUT SUPPOSE THERE WERE DISCREPANCIES BETWEEN TWO TIMEKEEPERS AND SCIENCE *COULDN'T* EXPLAIN IT BY APPEALING TO DIFFERING FORCES ON ONE AND NOT THE OTHER?

This is the possibility that occurred to Dirac and Milne. They imagined that the best electromagnetic clock, say our NIST F-1, might be initially synchronized with the best gravitational clock, say radio pulses from some nebula, and yet the two eventually come to disagree without any explanation. As far as we know, this is perfectly possible.

If this happened, it would probably spoil any hope of an eventually unified physics, where all of these phenomena are unified at some deep level. Fortunately, Nature has been kind to us. There aren't any discrepancies between our gravitational and electromagnetic clocks that don't already have a natural explanation at hand.

The Nature of Time: Relative and Non-Relative

In ordinary conversation we make reference to time in essentially two different ways. Sometimes we say of events that they are in the past, present or future: "The American War of Independence happened in the past", "My death is in the future", and so on. And sometimes we simply refer to events as being earlier or later than other events: "The American Revolution happened earlier than the French Revolution", "My death will happen later than my birth", and so on.

It takes a little bit of thought to see that these two ways of speaking are in fact different. Using the second way, claims about the time of one event are always made relative to other times of other events.

The American Revolution is not just earlier. That doesn't make sense. It's like saying "the pencil is on the right" without mentioning anything that it's on the right of. Rather, the American Revolution of 1776 is earlier with respect to events later than it, e.g., the French Revolution of 1789. And it is present with respect to those events simultaneous with it, e.g., the founding of San Francisco in 1776. And it is later than those events earlier than it, e.g., the accession of Louis XVI in 1774. In this way of speaking, whether some event is in the past, present or future is just a matter of what other time you're comparing it to. Times are always *relative* to other times.

Tenseless and Tensed Theories of Time

By contrast, in the first way of speaking, times are *not relative* to other times. One says that some event is in the past or future the way one might say that some jewels either are or aren't in a safe. Just as jewels cannot be **in** the safe with respect to one thing but **not in** the safe with respect to another – they're either in it or not – so too events are either past, present or future. Not past, present or future with respect to some other time – just past, present or future, full stop!

THE AMERICAN REVOLUTION ALREADY HAPPENED, SO NOW IT'S LOCKED IN THE PAST

SINCE IT HAS HAPPENED AND IS IN THE PAST, WE DON'T **NEED** TO ADD FURTHER THAT IT IS IN THE PAST WITH RESPECT TO THE YEAR 1920

There are also theories of time corresponding to these two modes of speaking. One theory, the **tenseless** theory of time (also called the static theory or the block universe) holds that time is much like space; the other, the **tensed** theory of time, holds that time flows or becomes, that it is a dynamic (changing) entity unlike space. Although hints of this distinction can be found in Aristotle, Augustine and elsewhere, the issue is a thoroughly modern one arising from debates among the philosophers John Ellis McTaggart, Bertrand Russell and C.D. Broad in the first quarter of the 20th century.

Tensed Time

The tensed theory of time probably best corresponds with one's intuitive idea of time, or the idea of time shared with the proverbial "man in the street". On this theory, the future is unreal. The event corresponding to what you will do after you read this sentence does not exist. The future is unsettled and ripe with possibility. As time passes, the world "chooses" one path from among all the available ones. The past is set and the present is that instantaneous point where the past and future meet. The world, in this picture, has the structure of a branching tree . . .

This theory corresponds to our idea that "what's done is done", that the past cannot be changed, and that the future can be changed because it is "open".

To take an example, consider how aspects of the world changed during Socrates' life. **Socrates** was born in 470 BC. Then as a baby, his not-yet-existent future was full of possibility . . .

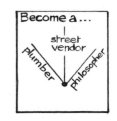

Socrates later served as an infantryman with conspicuous bravery in the Peloponnesian War (431–404 BC), and later devoted his life to teaching in the public spaces of Athens. His choices determined a unique "settled" past. After his trial in 399 BC, he was condemned to death by drinking poison hemlock.

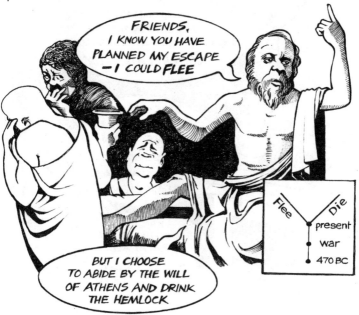

At the instant of his death, his entire life was settled and in the past.

There are in fact many versions of the tensed theory of time. In the version sketched above (the traditional one), the past is real and the present moves up the "tree" turning the unreal future into the real past. The English philosopher, **C.D. Broad** (1887–1971), perhaps best articulated this view. According to the version known as presentism, associated with the philosopher **A.N. Prior** (1914–69), the past and future are unreal and only the present is real. And on another version, the past, present and future are real, but the present somehow moves. In what follows we will stick with the traditional theory.

Although theories of science before 1900 did not entail this picture of time, they were compatible with it. Moreover, there is some reason to think that this is the picture most people had in mind.

Tenseless Time

The tenseless theory of time is less common-sensical but probably preferred by most (but definitely not all) philosophers and scientists. The main idea is that there is no becoming, branching or passage, that it's fine to represent time the way we represent space.

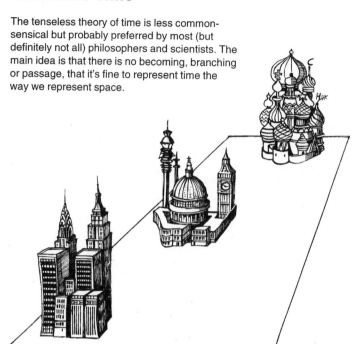

Just as New York, London and Moscow all exist but not at the same place, so the past, present and future all exist but not at the same times.
In this view the events of your birth, your reading of this sentence, and your death are equally real and on a par.

To understand this theory, let's begin with the idea of a dimension. Time, in this approach, is a kind of **fourth dimension**. As we'll see, there is nothing spooky about the fourth dimension – all it really means is that events can be represented with four numbers.

Representing Dimensions

A perfect mathematical point is zero-dimensional. It has no height, width or length.

•

Now drag the point two inches to the right, say, and pretend that it leaves ink behind.

───────────────

You have now drawn a one-dimensional line. It has length across the page, but neither width into the book nor (ideally) any height up or down the page. If we now take the line and drag it up two inches, we have a two-dimensional entity, a **plane**.

You can think of a plane as the thinnest possible sheet of paper with absolutely no thickness at all. If we pull the plane two inches out of the page, we have a solid **three-dimensional cube**. Our world is entirely populated with objects that are spatially extended in three dimensions.

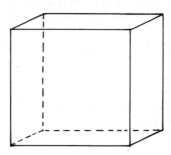

The Fourth or Time Dimension

Good. Let's press on. Although we can't visualize it in our mind's eye, it is easy to repeat the above procedure as many times as we like. In each case we take an object with number of dimensions n and imagine it being stretched in a direction perpendicular to all other directions to obtain the higher, n + 1, dimensional object. Let's do that one more time. And let's call that new dimension **time**.

Unfortunately, we can't draw this, even though we can easily understand it, as we'll now see.

You are probably familiar with the practice of drawing little stick people on the bottom corner of a pad of paper and then flipping the pages rapidly. It will look as if the little stick people are moving. Drop the page-flipping, and that is the idea here. The world is like the pad: a bunch of spatial entities located at spatial points (places on each sheet) at different times (different sheets).

Diagrams of Space and Time

Snap your fingers. Let this snap mark out an instant of time. At this instant all the objects in the world have a definite spatial location: your hands are where they are, the people on the plane overhead are a certain distance and direction from your hands, and so on. Now snap your fingers again. Now your hands have moved slightly (hand moving, earth spinning), as has the plane.

These snaps are marking out different places in space and time.

We can picture this if we agree to represent three-dimensional spatial objects as two-dimensional or even one-dimensional instead. Here the "height" in our diagram is not spatial height but rather time. Your hand and the plane (forget about the rest of the world for now) each trace out a "worldline" in our diagram, and your snaps pick out particular times on these worldlines.

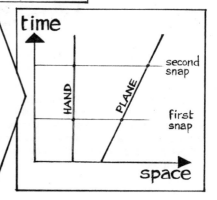

The motion of the plane away from your hand is represented by the increasing spatial distance (as measured by the horizontal axis) between the two as you travel "up" the diagram (that is, as time progresses, measured by the vertical axis).

Taking the easiest example, a rock sitting still (ignoring the motion of the earth) would look like . . .

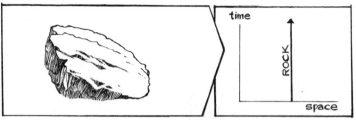

Two billiard balls colliding would look like this . . .

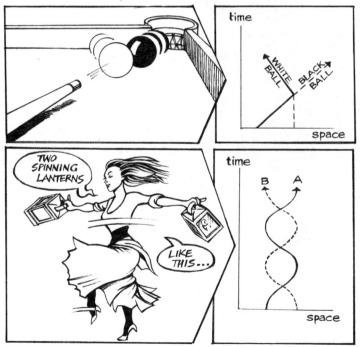

Picture of a "Tenseless" Life

In this tenseless or "block" view, the past, present and future all exist. Thus your life might look like this . . .

When describing this theory we must be careful, for we should not claim that the past and future exist *now*. "Now" has no privileged position according to this theory. "Now" is merely the time you say the word "now". If you say, "now this book is describing a strange theory of time", it refers to the events concurrent with this utterance. If you say it earlier, it refers to those earlier events.

The Now and the Here

"Now" functions just like the spatial sense of "here". "I'm here" refers to the location you're in when you utter the phrase – in New York it refers to New York, in London to London and on the moon to the moon. In this view, in sharp contrast to the tensed theory, being past, present or future is always **relative** to where you are on the space–time block. This theory initially strikes some as crazy.

We happily admit that Boston, London and Moscow are all equally real, even though you can't see one from another. The explanation is that they are at different places. Same here. The events of your birth and death exist but at different times.

The Problem of Motion and Change

People sometimes also object that there is no motion or change according to this tenseless theory. There is one sense in which this is true and another in which it is false. The sense in which it is false is if change means *only* the having of different properties at different times (and motion the having of different spatial locations at different times).

PLAUSIBLY, ALL IT MEANS FOR DR JEKYLL TO TURN INTO MR HYDE, IS FOR JEKYLL TO BE A NORMAL HUMAN BEING AT ONE TIME...

... THEN, AT ANOTHER TO BE MORE MONSTROUS

AND, AT ANOTHER EVEN MORE MONSTROUS AND SO ON

This is the way the philosopher and Nobel laureate **Bertrand Russell** (1872–1970) conceived of change. In this sense, there definitely is change and motion in this theory.

Your tea cup initially has the property of being warm and then at another time, less warm, at another time still less warm, and so on. The moon has the property of being at one place at one time, at another at another time, and so on.

The former describes the change from hot to cold, the latter the kind of change known as motion. All there is to change in this view is a *three-dimensional* object possessing different properties at different times, and this can certainly occur in the tenseless theory. Motion and change are already encoded in the four dimensions.

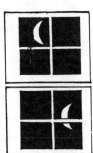

Of course, in the tenseless theory, the temporal relations among *all four-dimensional* objects are perfectly fixed, once and for all. Looked at four-dimensionally, then, there is no change. The aliens in Kurt Vonnegut's novel *Slaughterhouse-Five* (1969) could see four-dimensional objects, not merely three-dimensional ones. They thus saw the history of Earth, past, present and future all laid out, fixed and unchanging. Since there is no **fifth** dimension with respect to which the four-dimensional objects can change, Russellian change cannot occur at this level. Nor should it, says the *detenser* (the opponent of tenses).

The property of being 100 degrees C at time t is not only not located at time t + 1, but it has also fallen into the Past. This "something extra" is needed, they say, to distinguish real temporal change from mere variation (for example, the variation in colour on a flag). In this more robust sense of change, there is no change in the tenseless theory.

McTaggart's Argument

In "The Unreality of Time" (1908), the Scottish philosopher **J. E. McTaggart** (1866–1925) presented an argument that is now famous among philosophers. The argument concludes that time *doesn't exist*, or better, that nothing worth calling time exists. There is something in McTaggart's argument for tensers and detensers alike. The part tensers like is his claim that *genuine* change requires the tensed theory of time to be true. For McTaggart, it's not enough to say that the teacup is hot at time t but not hot at time t*, where t* is later than t.

So McTaggart thinks that the tensed theory of time best fits our experience.

The funny thing is, he states, although tenses are best, they're actually *incoherent*. Unsurprisingly, this is the part of the argument detensers favour. The argument is deceptively simple, arising from merely two claims made by the tensed theory.

1 PAST, PRESENT AND FUTURE ARE INCOMPATIBLE PROPERTIES.

IF AN EVENT IS PAST, FOR EXAMPLE, IT CANNOT BE PRESENT

2 BUT EVERY EVENT HAS ALL THREE OF THESE PROPERTIES. SOCRATES' DEATH WAS ONCE FUTURE, THEN PRESENT AND IS NOW PAST

CLAIMS 1 AND 2 ARE *BOTH* TRUE ACCORDING TO THE *TENSED THEORY* OF TIME, BUT THEY ARE *LOGICALLY INCOMPATIBLE*

Therefore, the tensed theory of time must be false. But since he thinks that it's the best account of time and change available, his reasoning leads him to the staggering conclusion that time is unreal!

Avoiding McTaggart's Trap

There are two general strategies for avoiding this conclusion. Detensers simply deny that the tenseless theory of time is an inadequate account of our experience of change.

Socrates' death is past **in 2000 AD**, present (as best we know) **in 399 BC**, and future **in 500 BC**, not past, present and future all at once, as claim 2 says.

By itself, this answer is no good (which is why McTaggart's argument has stuck around). To state that "Socrates' death is past in 2000 AD" is to state a tenseless fact.

IT'S ALWAYS TRUE, REGARDLESS OF WHAT TIME IT IS... AND IT'S RELATIONAL— THE DEATH IS PAST **RELATIVE TO** 2000 AD

HENCE, **TENSERS** CANNOT CLAIM THAT THIS IS WHAT THEY MEAN

If they remove the inconsistency between claims 1 and 2 by adopting the tenseless theory, they've given up their theory. The open problem, then, is for tensers to see whether they can really make sense of tenses in a non-tenseless way.

How Fast Does Time Flow?

In a famous essay entitled "The Myth of Passage" (1951) the American philosopher **D. C. Williams** launched a powerful attack on the tensed theory of time. Williams asks whether there is really need to posit tenses. Thinking the alleged needs are all based on confused understanding of the block universe (such as that it can't represent motion), he claims (with characteristic flair) that the tenseless theory is "the logical account of the events par excellence, the teeth by which the jaws of the intellect grip the flesh of occurrence".

This paper contains a famous objection to tenses, made originally by C. D. Broad but generalized by Williams. The objection arises from the question: "How fast does time flow?"

So, if the Now moves, it must move with respect to time. But the Now *is* time. So is it moving with respect to itself? No, says Williams, that doesn't make sense. If it's moving, it must be moving with respect to a second time – the second one being the time with respect to which the first one moves.

Tensers sometimes reply to the question "How fast does time move?" with "One second per second." But notice that this doesn't answer the question, for we can simply cross out the seconds and we're left with an answer of one. One what? Just one. This fails to give a proper solution.

The best answer, perhaps, is for tensers to deny that the Now moves with respect to time at all. C. D. Broad held such a view:

THE CHANGE THAT THE **NOW** UNDERGOES IS A UNIQUE FUNDAMENTAL PROCESS NOT ANALYSABLE IN TERMS OF MORE FAMILIAR NOTIONS OF CHANGE

THIS ANSWER LEAVES THE ISSUE STILL A BIT MYSTERIOUS BUT IT REMOVES THE NEED TO START POSITING EXTRA DIMENSIONS OF TIME

Finally, let's mention that the tenseless interpretation of time is fully compatible with the times of Leibniz, Newton, Poincaré, and (as we'll see) Einstein. Whether the tensed theory can say the same regarding Einstein is a topic we'll confront later.

Galilean Relativity

The Italian scientist **Galileo Galilei** (1564–1642) best articulated a familiar but interesting fact about the world. The laws of physics are indifferent to the speed at which a body maintains straight uniform movement.

Using the example of a boat in calm seas travelling at a constant rate in a straight line, Galileo pointed out that Nature doesn't care whether you are doing your scientific experiments on this boat going uniformly at 5 mph, 10 mph or not going at all (0 mph). You get the same phenomena, the same results.

Indeed, if the motion of the boat was perfectly uniform, and you were in a cabin with no windows, you wouldn't be able to tell whether the boat was moving or not. This remarkable fact – that Nature doesn't care whether you do your physics while moving uniformly or while still – is called **Galilean relativity**.

Frames of Reference

Another way to state this Galilean principle is that Nature doesn't distinguish between non-accelerating, or "inertial", **frames of reference**. A frame of reference is simply a bunch of objects that are not moving relative to one another. The above boat defines a frame of reference. The boards, tables, masts, etc. are not moving appreciably with respect to one another.

There is an infinity of frames of reference and, of course, they can move with respect to each other.

The speed of things depends on what frame of reference you're in. If you're driving a car at 50 mph, then the roadside passes by you at 50 mph. Someone standing on the roadside will also see you going at 50 mph.

Einstein's Relativity

In 1905, our understanding of the world changed. A young **Albert Einstein** (1879–1955) enjoyed his *annus mirabilis* (year of miracles): while working in a patent office, he published three articles, each of them revolutionary in physics. One of them introduced his most famous theory, **special relativity**. Over the next few pages, we'll briefly explain the consequences that special relativity has for time. Later, when discussing time travel, we'll also examine Einstein's 1917 masterpiece, **general relativity**.

In 1905, Einstein assumed – and was vindicated by experiment – that light looks the same in every *non-accelerating* frame of reference.

This is strange. If we're both on a train and I throw a ball to you, then it may travel at, say, 20 mph relative to our frame of reference. But if you're standing at a level crossing waiting for the train to pass, and the train is going at 100 mph, and I throw the ball to you in the same manner as before, then the ball will go forward at 80 mph, which would make it much more difficult to catch!

BUT IF I 'THROW' LIGHT AT YOU—BY TURNING ON THIS LAMP—THEN THIS BASIC EFFECT WON'T OCCUR

THE LIGHT REACHES ME AT THE SAME SPEED WHETHER THE LAMP IS STATIONARY OR MOVING AWAY FROM ME AT 10,000 MPH

Simultaneity is Relative to the Observer

Merely from Galileo's observation that the same physical laws hold in non-accelerating frames, and the bold supposition that light travels always with the same velocity, strange things follow for time.

Consider again a train. Suppose I go midway in a carriage with a lamp. I turn on the lamp. From the perspective of someone in the carriage, the light should arrive at both exits at the same time. That is, the event A = light arrives at the front door is *simultaneous* with the event B = light arrives at the back door.

HOW DOES IT LOOK **FOR YOU** AS THE TRAIN RACES AWAY?

IT LOOKS AS IF THE BACK DOOR IS **CHASING** THE LIGHT, WHEREAS THE FRONT DOOR IS **RUNNING AWAY** FROM IT

For the bystander, event B comes *before* event A. What is simultaneous for the passenger is not simultaneous for the bystander.

Newtonians, of course, would not suggest anything so radical. They would want to say that the light travels at different speeds, depending on your frame of reference.

BUT THIS IS NO LONGER AN OPTION, SINCE WE ASSUME THAT LIGHT TRAVELS AT THE **SAME SPEED** IN **ALL** FRAMES OF REFERENCE

The idea that simultaneity depends on who is doing the observing is a profound consequence of Einstein's assumptions: Galilean relativity and the constancy of the speed of light. And it has important implications for the nature of space and time.

Earlier, when explaining Newton's theory of time and the block universe, we tacitly assumed that simultaneity is independent of particular observers. There was one unique global time. If I snap my fingers and you snap your fingers, these events are either simultaneous or not – independent of how anyone or we are moving.

There isn't time as we typically think of it. As Einstein's teacher **Hermann Minkowski** (1864–1909) put it, there isn't even time any more . . .

"Henceforth space by itself, and time by itself, are doomed to fade away´ into mere shadows, and only a kind of union of the two will preserve an independent reality."

The Spacetime Event

There is a single entity, **spacetime**, not space and time. And it is sliced up into space and time in different ways, depending on the observer. Suppose we think of two space shuttles drifting towards each other.

Lightcones

Because light has no mass, travels faster than anything else – so fast that it has practically the same speed in all reference frames, it is in a special position to tell us about the structure of spacetime in relativity. Following light will help us to picture spacetime. To better understand special and general relativity, we therefore need to understand **lightcones**.

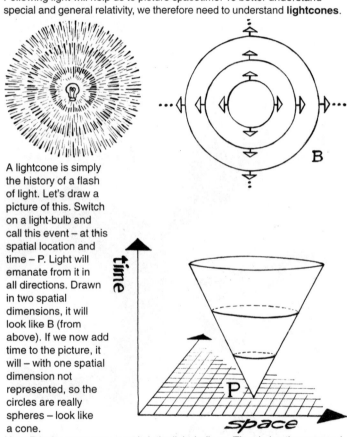

A lightcone is simply the history of a flash of light. Let's draw a picture of this. Switch on a light-bulb and call this event – at this spatial location and time – P. Light will emanate from it in all directions. Drawn in two spatial dimensions, it will look like B (from above). If we now add time to the picture, it will – with one spatial dimension not represented, so the circles are really spheres – look like a cone.

Here P is the moment you switch the light-bulb on. The circles then expand up the diagram as the light travels further and further away from the light-bulb with increasing time. The paths of the light rays form a cone – thus the term lightcone. This particular cone is called the *future lightcone*.

The point in the diagram – that event of switching on the light-bulb – also has a *past lightcone* that we may draw.

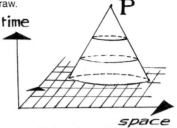

The past lightcone represents all the rays of light in the universe that could make it to the point in spacetime, P. Light goes fast, but it still goes at only a finite speed. The light from every light-bulb won't be able to make it to every event. Thus the light from the bulb turned on at point Q can't make it to point P. Light rays in these diagrams always travel away from points in spacetime at 45-degree angles.

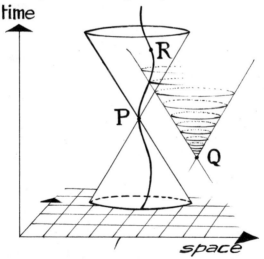

Thus the observer at P can't see the light-bulb at Q switched on, but if she waits a while, until R, then she will be able to see it. This is in fact very familiar. When looking into the night sky, your eye is receiving light rays that have travelled from stars for millions of years. The previous night, you couldn't have seen those very light rays. They were outside your past lightcone, but today they are in it.

Time and Observer Dependency

One important fact about lightcones is that they represent the limits of which events can affect one another. Nothing goes faster than light, so anything that will influence you must be travelling either on the lightcone itself (if it's light) or within the lightcone (if it's going slower than light). The same goes for anywhere you hope to go or influence.

Now, we've drawn our picture with one time, but that was merely for the sake of simplicity. The march of time is observer-dependent, to a certain extent. Within one particular observer's lightcone, the order of events is definite. But another observer, moving relative to our first observer, will disagree with the first as to what events are simultaneous with event P.

If the three observers met up later in the future, they might quarrel over the correct order of events. "Surely," they would each think, "the others must be confused: there really is a fact of the matter about whether A happened before, with or after B?" Amazingly, if Einstein is right, the answer is "No, they are each right."

There is no one unique time in a relativistic universe. Rather, there is a plethora of times, one for every inertial frame of reference, and they are all equally legitimate. There is also your so-called proper time. The proper time measures how much time has passed on your particular path in spacetime. Depending on your travels, then, your proper time may be very different than that of some other travellers.

Earlier, we discussed a father and son largely agreeing about the order of events. Relativity says there is no *absolute* ordering that will be true for all people at all places at all times (though the *relative* ordering for each person at each point in spacetime remains objective).

Slower or faster, some on planes and others stuck at their desks, we're all travelling pretty slowly with respect to each other from the perspective of relativity. To get serious disagreement about the ordering of events, you need to go *seriously* faster or slower than someone else.

Relativity and Tenses

Before moving on to time travel, let's pause to examine an argument by various philosophers which claims that special relativity makes the common-sense tensed theory of time impossible. The idea, often associated with the American philosopher **Hilary Putnam** (b. 1926), is simple. The present, according to the tensed theory, turns the unreal future into something real, from something open to something fixed. In the Newtonian picture, everyone agreed on which events were present.

IN RELATIVITY WHAT IS **PRESENT** DEPENDS ON THE OBSERVER

WHAT IS PRESENT FOR OBSERVER 1 IS **NOT** PRESENT FOR OBSERVER 2

WORSE, EVENT B IS IN THE FUTURE OF OBSERVER 2 AND THUS **UNREAL** & INDETERMINATE

...WHEREAS IT'S IN THE PAST OF OBSERVER 3, AND SO **REAL & DETERMINATE**

Well, which is it? Events are either real or not. Surely it's ridiculous to state that whether an event exists depends on how fast you're moving. Yet this seems to be what we have to say if we merge the tensed theory with time in special relativity.

Consequently, it seems that even if the philosophical objections to the tensed theory haven't killed it already, then special relativity would finish it off. Yet, there are still defenders of the tensed theory of time who believe they can explain away this conflict. And there are also those who favour an alternative interpretation of relativity by the Dutch physicist and 1902 Nobel prize winner **H.A. Lorentz** (1853–1928).

THE DISTINCTION BETWEEN RESTING AND MOVING WITH UNIFORM SPEED IS MEANINGLESS BECAUSE WE CAN'T DETECT THE DIFFERENCE WITH AN EXPERIMENT. YET THIS ASSUMPTION IS AT THE HEART OF THE THEORY

But as Lorentz saw it, a state of "real" rest exists even though it cannot be observed by experiment. True, we can't tell experimentally which is the right way to slice spacetime into space and time (which is what makes Einstein's theory possible), but that doesn't mean that there isn't one correct way of slicing the world into space and time.

So the Lorentzian theory does not have the profound consequences for time that relativity has.

In any case, defenders of tenses have a lot of work to do. Not only do they have to answer McTaggart and Williams, but now they must contend with Einstein too.

Does Logic Allow Time Travel?

Now that we understand the tenseless theory of time and special relativity, we're ready to tackle the question of time travel. The question is natural to the tenseless conception of time, for if time is like space, why can't we visit other times just as we visit other places? Just as we book a vacation to the Greek Isles, why can't we book a vacation to Ancient Greece? We wouldn't have to worry about crowds on the beaches, after all.

It is no accident that the most famous time travel story of all, H.G. Wells's *The Time Machine*, begins with a dialogue between the time traveller and others about the tenseless theory of time. Though not recognized as the tenseless theory back in 1895 when his book came out, Wells definitely has the tenseless theory in mind. He says time is simply the fourth dimension and compares a time machine to a hot-air balloon.

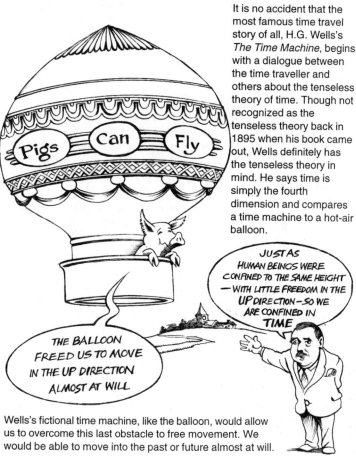

Pigs Can Fly

JUST AS HUMAN BEINGS WERE CONFINED TO THE SAME HEIGHT — WITH LITTLE FREEDOM IN THE UP DIRECTION — SO WE ARE CONFINED IN TIME

THE BALLOON FREED US TO MOVE IN THE UP DIRECTION ALMOST AT WILL

Wells's fictional time machine, like the balloon, would allow us to overcome this last obstacle to free movement. We would be able to move into the past or future almost at will.

Time travel is *logically* possible, contrary to what many have asserted. That is, the idea of time travel is not a logical contradiction. Logical contradictions are statements such as "The tall man is not tall" and "I went to the store and didn't go to the store". They describe impossible scenarios. Many who have thought about time travel have concluded that it really doesn't make sense, but we will show that it is perfectly coherent. Of course, this does not show that it's *physically* possible.

Why would someone think time travel logically impossible? There are a number of reasons, but they all concern the oddities that might result if time travel were possible. Let's consider a few consequences, beginning with tame ones and then working our way up to true weirdness.

The Logic of Impossibility

If I had the ability to send things back in time, I could transmit a letter to you, asking for your phone number, so that you received the letter at a time before I sent it.

You might then send me your phone number before I write the letter.

If that happened I would have no need to write the letter, since I already would have your phone number.

But then I wouldn't have it if I hadn't written and sent the letter . . . and so on.

This is no doubt weird and awkward, but it is far from a logical contradiction, which is what we care about right now.

They might have forgotten about the received letter with the phone number, or maybe they just want another letter for whatever reason. People are strange; that doesn't threaten logic.

The Book That No One Wrote

Let's go a step closer to a contradiction. In the year 2000, at a bookstore on Charing Cross Road in London, I buy Charles Dickens's *A Tale of Two Cities*. I use my time machine to deliver the book to Dickens's door in 1855, a few years before we think he wrote it.

I THEN HAVE ACCESS TO THE ENTIRE FINISHED BOOK BEFORE I'VE WRITTEN A WORD

He then copies it and publishes it in serial form, beginning in the year 1859.

The story is widely read, it becomes famous, he dies, and publishers collect it together into a book and sell it for years – and ultimately to me in 2000.

Strange: he had a surprisingly easy time with this book, for he never had to write it at all. But if he didn't write it, who did? No one did! True, every copy of the story in every book everywhere in the world was actually produced by a hand or machine. But if this was the way the world had worked out – and it could have been if there is time travel – then the information and ideas in *A Tale of Two Cities* were never created by anyone.

The Causal Loop

The storyline of *A Tale of Two Cities* simply exists timelessly, in what philosophers call a *causal loop*. Consider three events in this loop: A, Dickens copying the manuscript; B, Dickens performing readings of it in public; and C, publishers binding my copy in 1999. Each stage in this scenario partially caused the next. A is a partial cause of B, and B is a partial cause of C.

NORMALLY THIS IS HOW THINGS TRANSPIRE: THE CAUSES ARE EARLIER THAN THE EFFECTS. HOWEVER, TIME TRAVEL ALLOWS ANOTHER STORY, WHERE C IS A PARTIAL CAUSE OF A EVEN THOUGH A IS EARLIER THAN C

There is no logical inconsistency here. Nothing both happens and doesn't happen. The four-dimensional loop just exists, and despite its oddity it's perfectly conceptually coherent. Perhaps things like this don't happen in our world; perhaps such a loop even violates the laws of Nature. But it doesn't violate the laws of logic.

Incidentally, if you have seen the Hollywood film *Terminator*, you will have encountered a seemingly consistent causal loop, perhaps without realizing it. In the film, evil computers and robots take over the world and enslave the human race.

A rebel human being poses a serious threat to them, however, so they invent a killing machine, the Terminator (Arnold Schwarzenegger), and send it back in time to kill the rebel's mother before he is born. The Terminator loses the battle . . .

BUT THE CRUCIAL THING IS THAT PARTS OF MY BODY STICK AROUND IN THE PAST AND IT'S THE TECHNOLOGY FROM THESE VERY PARTS THAT SPAWNS THE DARK EVIL ROBOT FUTURE

If they hadn't sent the Terminator back, they wouldn't have come to be in the first place! Yet in the story, they *are* around, therefore the Terminator must already have arrived back in the past to ensure that they would be. Again, just as the information of the plot of *A Tale of Two Cities* was never created in our story, nor is the technological information needed for the evil robot society ever invented.

A Logical Contradiction of Time Travel

Okay, now here is an apparent consequence of time travel that does violate the laws of logic. Imagine that for some reason you hate your life and want to kill yourself. Being a tidy person, you want to erase your *entire* existence, not merely from now onwards. You want to extinguish every trace of your wretched life. So you hatch a plan.

According to one "history" of the universe, you existed (say) as an eight-year-old in 1985. According to another "history", you were killed as a baby and there was no eight-year-old you in 1985. But there is only one world, one history.

So, if there aren't any alternative "parallel" universes or histories or anything funny like that, it seems that time travel would allow logical paradoxes.

Logical Contradictions Cannot Occur

This possibility of paradoxes is what worries the Cambridge physicist **Stephen Hawking** (b. 1942). He says that he wants to show that the world is safe for historians (more on this later). However, if Hawking had noticed a paper by the American philosopher **David Lewis** (b. 1941), then perhaps he wouldn't worry so much about time travel allowing logical paradoxes.

There are two points to make to see that time travel is possible. First, logical contradictions just can't happen, so nothing can "allow" them to occur. The philosopher Robert Weingard puts the point like this . . .

SINCE LINCOLN **WAS** ASSASSINATED, IF I GO BACK IN TIME, I WON'T STOP THE ASSASSINATION BECAUSE, OF COURSE, **I DIDN'T!**

And in general, if you do go back in time, since your backward trip has already taken place (relative to your time of departure), you will not create a contradiction on your trip because you didn't.

PUT DIFFERENTLY,
SINCE CONTRADICTIONS
ARE IMPOSSIBLE, THE ARGUMENT
SHOWS THAT THE ONLY TIME TRAVEL
TRIPS THAT WILL EXIST ARE
ONES THAT ARE NOT
CONTRADICTORY

And so *there are* time travel scenarios that are perfectly consistent.

Personal Time

Following Lewis, let's speak of a person's "personal time". Personal time is an ordering parameter that we define from a person's physical and mental processes. Imagine that people carry a wristwatch on them measuring this time.

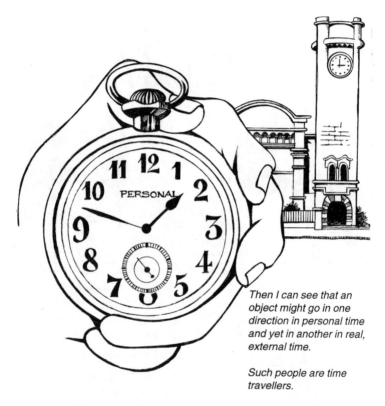

Then I can see that an object might go in one direction in personal time and yet in another in real, external time.

Such people are time travellers.

Apparently we aren't time travellers because our personal time matches the real external time. But logically speaking, there is nothing to prevent one's personal time from not matching the ordering of external time. Let's imagine an example of this.

Dying at a Time Earlier Than Your Birth

Suppose, for instance, that you are born in 1980 and visit London in 2010. While in London you go to make a phone call but step into Dr Who's phone-box time machine (the Tardis) by accident. While fiddling with some knobs, you foolishly send yourself back millions of years to the Jurassic period. After stepping out of the machine you panic, hide, and eventually squeeze out five desperate years before being impaled by a massive triceratops.

Tragically, you have died before you were born.
That is, in external "real" time you have died millions of years before you were born, but in your personal time you died in your 35th year.

There is nothing logically incoherent about this Jurassic scenario. You may have died before you were born, in one sense – but, in another sense (in terms of personal time), you died after you were born.

Nothing in logic or the tenseless theory says that you can't die at a time earlier than you were born, just as nothing says you can't die in the Northern hemisphere, even if you were born in the Southern.

The important point to remember in the Jurassic story is that if this happens, then from our point of view you already were in the past, so the future had better be compatible with this. The world only "runs" once.

Future Compatibility

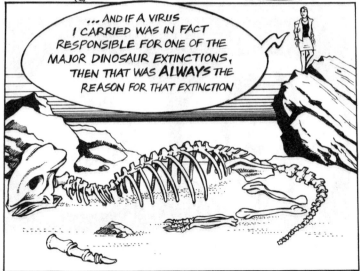

Can We Change the Past?

This raises the most popular question about time travel. Can we change the past if we go back in time? The answer depends on what you mean by "change the past". If by "change" you mean making an existing event not exist, then you can't. The event would exist and not exist, which is impossible. *In this sense* not only can't we change the past according to the tenseless view, but we can't change the present or future either. We can't make contradictions true!

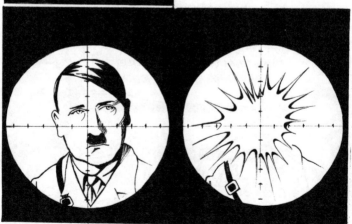

So, if the event of World War II exists, we can't make it not exist when we go back in time.

Can We Affect the Past?

And if right now you're poor, nothing you do in the past can change this fact. But that doesn't mean we can't affect the past *in the more ordinary sense*. If we can go back in time to a certain point, we can affect the past after that point the same way that in the present we affect the future. But if we did go back to a time earlier than the present, then it "already happened" and it would not change the present that we know.

Two Sorts of Time Travel Stories

There are consistent time travel stories and inconsistent ones. Consistent stories don't have events "disappearing", whereas inconsistent ones do. In many stories, for instance the US TV series *Quantum Leap* and certain episodes of *Star Trek*, the whole idea is to change events that *already happened*. According to the tenseless theory, this intention is fine so long as events aren't actually altered. But in these shows there are always two possible futures – one wherein things "go wrong" and one wherein things "go right".

ACCORDING TO THE TENSELESS THEORY OF TIME, THIS IS SIMPLY IMPOSSIBLE BECAUSE THERE IS **ONLY ONE** ACTUAL FUTURE

By contrast, one children's story involves a time traveller going back to prehistoric times, giving the cavemen certain information, and coming back.

Meanwhile in the present, cave paintings exist that picture a person, who looks exactly like the traveller, teaching Stone Age people various things.

This is consistent: the traveller went back to the past, affected it, and the future matches this. It's what always happened.

Many other time travel stories are also consistent, including some very bizarre ones. Robert Heinlein's *All You Zombies* (1959) is perhaps the weirdest consistent story. In it, an orphaned girl becomes her own mother and father by having a sex change operation and then travelling back in time.

Establishing the logical coherency of time travel is one thing, but it's another to say that it can actually happen. After all, it's logically possible that pigs fly. So we should now turn to the question of whether time travel is physically possible. Do the laws of physics allow one to travel back in time?

Does Physics Allow Time Travel?

To begin with, we should mention that, thanks to special relativity, there is a type of time travel already available to us. But this type of travel is not too exciting. This is travel making use of relativistic "time dilation". You may have heard of the famous "twin paradox". In it, one twin leaves the Earth in a rocket and returns in what seems to her to be only five years.

By going faster, the twin could have come back and discovered that hundreds of years on Earth had passed. In this sense, she can travel in time.

Moving Clocks Run Slower

The reason "time dilation" works is as follows. Snap your fingers . . . now snap them again. Call the first event A and the second one B. Using your wristwatch, you will measure the time interval between snaps to be a certain amount, T. Someone moving by you (who is not accelerating or decelerating) will, according to their wristwatch, measure the time interval between A and B to be an amount T*.

If the person moving by you is moving slowly compared to you, then γ is close to 1 and thus T almost equals T*. You therefore see A and B as happening with the same duration between them, say 5 seconds.

But if the person is moving very, very quickly relative to you, close to the speed of light, then γ is close to 0 and T can be very different from T*.

The clock of the twin in the rocket is running slow compared with the clock of the twin left behind. That is why the travelling twin can age only 5 years while the Earth-bound twin ages 30.

Small Savings in Time

Time dilation has been observed in a variety of experiments. Perhaps most impressive is the direct evidence provided by comparing atomic clocks in jets with those on the ground.

*This was shown in an experiment by **Joseph Hafele** and **Richard Keating** in 1972.*

They found that flying clocks placed in a high-speed jet lost about 59 nanoseconds while flying eastwards . . .

. . . and gained about 273 nanoseconds while flying westwards. The discrepancy is due to the Earth's rotation.

The reason this type of "time travel" is boring is that, first, you don't get much return for your money. If you have unlimited funds and keep travelling on a supersonic jet around the world eastwards for 25 years, then, on landing, you would find that at best your clock is a few seconds behind the clocks on the ground.

The space shuttle would get you a little more time, but clearly a few seconds isn't worth the bother. Also, this time travel isn't the kind that one dreams about.

Relative to your own personal impression of temporal progression, nothing changes. In any case, it's one-way time travel – you can't get back. This may be disappointing, as you won't be able to return to tell about your exploits.

General Relativity and Four-dimensional Curvature

General relativity, Einstein's greatest triumph (discovered in 1917), does allow for more exotic time travel. To see this possibility at a simple level, fortunately we need not learn all of general relativity. We can instead reflect on one of the principal conceptual advances of general relativity; namely, the idea that four-dimensional spacetime can be curved.

We are familiar with curvature in everyday life. A one-dimensional line might be curved, for instance, like this . . .

Or a two-dimensional surface might be curved into a ball . . .

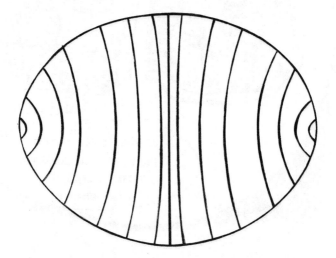

In each of these cases, we think of the object as being curved in the direction of a higher dimension. The curve of the one-dimensional line, for instance, is in the two-dimensional plane (the page).

Does this mean that for a four-dimensional object to be curved there must be a fifth dimension into which the four-dimensional object curves? No, there is a perfectly well-defined way of thinking about curvature as intrinsic to the object itself, without any reference to higher-dimensional spaces.

IMAGINE THAT YOU ARE SPATIALLY TWO-DIMENSIONAL AND TRYING TO DECIDE WHETHER YOU LIVE ON A FLAT SURFACE, OR A CURVED SURFACE LIKE THAT OF A BASKETBALL

WELL, WITH A COMPASS TO HELP GUIDE ME AS I WALK A STRAIGHT LINE I COULD DRAW A TRIANGLE ON THE SURFACE

$a+b+c = 180$

You know that the three angles of a triangle on a flat surface will add up to 180 degrees, no matter what kind of triangle you draw. So, you know you're on a flat surface.

Why We Don't Need a Fifth Dimension

But if you draw a triangle on a curved surface, say a basketball, the three angles will sum to more than 180 degrees. Take its north pole as one vertex and draw a line straight down to the equator. Turn the basketball 90 degrees and draw another line straight down to the equator. Now draw a straight line along the equator connecting these two lines.

The important point is that you, as a two-dimensional object, can do this experiment without any need whatsoever of higher dimensions. So we don't need five dimensions to make sense of the idea that four-dimensional spacetime is curved.

Spacetime Curvature

The evidence we have suggests that our universe is curved. This curvature explains the force of gravity. Light rays, for instance, travel in straight lines. But they have been observed – near the sun – to deflect slightly. General relativity explains this, very roughly, as the result of the sun distorting (curving) spacetime and causing the light to "fall" towards it . . .

MUCH AS A HEAVY BALL ON A RUBBER SHEET WILL BEND THE RUBBER AND SO AFFECT THE PATHS OF SMALLER BALLS NEARBY

Locally (in small regions), however, spacetime appears flat and special relativity is a good approximation.

General Relativity and Time Travel

Back to time travel. The point of the digression into curvature is that
general relativity allows really curved universes. And some of these
really curved universes allow time travel. So if general relativity tells us
what the laws of physics allow, then it tells us that the laws of physics
allow time travel. We'll return to this point in a moment, but let's first
see how this works.

Consider a flat spacetime of the sort we had been talking about until
now. For ease of visualization, just consider a world with one dimension
of time and one dimension of space, suppressing the other two spatial
dimensions. Think of this as simply a piece of paper.

This is a perfectly legitimate model of spacetime. It's not intrinsically curved – you didn't need to stretch or shrink the paper (triangles will still sum to 180 degrees). But by travelling in the future time direction, as normal, you will eventually return to your past.

Closed causal loops of the sort we discussed before (see pages 74 and 75) are even possible. This type of spacetime is allowed by general relativity. Time travel on "cylinder spacetime" is not what we usually mean by time travel, however. You never really travel backwards in time.

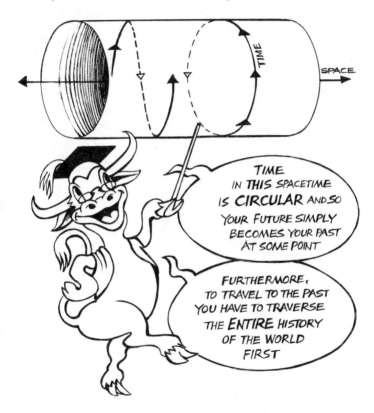

This isn't what you typically want when thinking of time travel.

Gödel's Rotating Universe

Allowing for spacetime curvature allows for very exotic spacetimes that permit a more interesting time travel. The most exotic such spacetime is undoubtedly the one discovered in 1949 by the great logician **Kurt Gödel** (1906–78). Gödel is famous for perhaps the most profound result in mathematical logic in the 20th century, the "Incompleteness Theorem".

LESS WELL-KNOWN IS THE FACT THAT I BEFRIENDED EINSTEIN AT PRINCETON

... AND AS A RESULT OF OUR DISCUSSIONS, GÖDEL FOUND NEW SOLUTIONS TO THE EQUATIONS OF GENERAL RELATIVITY

This is no mean feat. But for perhaps the greatest logician in the 20th century, solving these equations was probably like an ordinary person doing a crossword puzzle.

Each solution to the equations of general relativity describes a spacetime allowed by the laws of general relativity, and therefore, allowed by our laws of Nature. Gödel's solution is very strange. Our universe, we know, expands in all directions from every point. There is no centre.

And like our universe's expansion, the Gödel universe's rotation does not have a unique centre. Instead, from the perspective of any observer, all the matter in the universe is rotating.

Spacetime in a Rotating Universe

But rotation has a funny effect. If you take a paddle from a canoe or rowing boat and rotate it under water, the rotating paddle will cause the water to swirl all around it.

THE ROTATION **DRAGS** THE WATER AROUND THE PADDLE

THE SAME THING CAN HAPPEN TO SPACETIME

WHEN OBJECTS SPIN, THEY **DRAG SPACETIME** AROUND WITH THEM

The effect is usually a tiny one, so forget about noticeably dragging spacetime with a spinning top on your desk. But if all the mass in the universe is rotating about you, the effects of this dragging can be spectacular.

What "dragging" can do is so contort spacetime that the very futures of some events can get "tipped over". Let me explain. Earlier, when we discussed special relativity, we saw that every event has a future and past lightcone.

We assumed before that the lightcones of all events were oriented the same way.

The Effect of Spacetime Curvature

But curvature allows the lightcones to tip with respect to each other. Think of a person travelling on our rubber-sheet model of spacetime. Imagine the rubber sheet has a bunch of **X**s on it, all oriented the same way. The **X** represents the lightcone for that point and, as before, the **X** lines or "light rays" are all at 45-degree angles . . .

Again, place a heavy ball on the rubber sheet. The ball's bending of the sheet will have the effect of "tilting" the **X**s towards or away from each other, which will extend the limits of where the traveller can go. It is even possible for

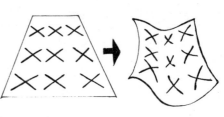

the **X**s to tip right over on their sides. If arranged the right way, we could draw a line on the sheet such that it came around back to its origin, even if throughout it only travels in the top part of the **X**s.

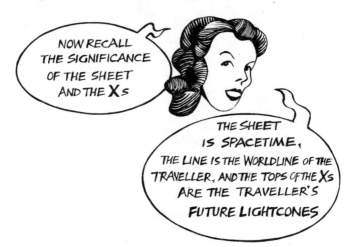

NOW RECALL THE SIGNIFICANCE OF THE SHEET AND THE **X**s

THE SHEET IS SPACETIME, THE LINE IS THE WORLDLINE OF THE TRAVELLER, AND THE TOPS OF THE **X**s ARE THE TRAVELLER'S **FUTURE LIGHTCONES**

Our little exercise shows that it is conceivable that spacetime be so curved that it tips the future lightcone into the past. And a traveller can make use of this tipping to come back to her past simply by travelling in her local future.

At a certain point, the lightcones are completely tipped over sideways. You could thus go far away, where the lightcones are tipped over, use this as a means of going "down" into your past, and then return to a point in the spacetime before the beginning of your journey!

Taub-NUT-Misner Spacetime

Taub-NUT-Misner spacetime is also a time-travel permitting solution to general relativity. It is also a kind of cylinder spacetime, like that formed from rolling up a sheet of paper, except that it is created through intrinsic curvature, not cutting and taping, and it "stands up" rather than "lies down." As you go "up" in time in Taub-NUT-Misner, the lightcones tip over.

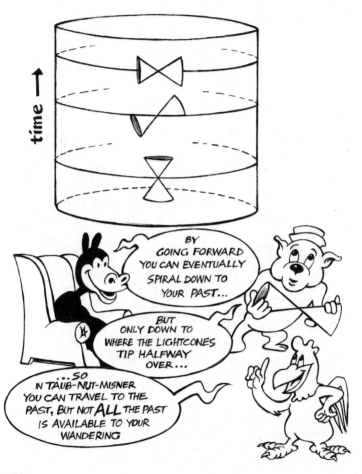

Gödel's Complete Spacetime Travel

One of the many amazing things about the Gödel spacetime is that through *every single point* in the spacetime there are physically possible paths that allow you to time-travel. And from each event you can reach any other event. As Gödel stated in 1949:

As you move outwards the time axis tilts right over

BY MAKING A ROUND TRIP IN A ROCKETSHIP IT IS POSSIBLE IN THESE WORLDS TO TRAVEL INTO ANY REGION OF THE PAST, PRESENT, AND FUTURE AND BACK AGAIN

EXACTLY AS IT IS POSSIBLE IN OTHER WORLDS TO TRAVEL TO DISTANT PARTS OF SPACE

Moreover, you need not traverse the entire course of world history to go back in time. You just need to go for a little "detour" first and then you can go anywhere (the further into the past or future, the bigger the detour).

Is Gödelian Time Travel Possible?

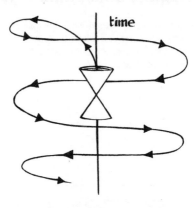

time

Of course, time travel in the Gödel universe will cost you. The philosopher **David Malament** (b. 1947) has worked out the energy requirements of travelling back in time in a Gödel universe, and he found that it is effectively technologically impossible to do it. So even if our world is Gödelian – and it doesn't look like it is because it isn't expanding like ours and ours isn't rotating like it – practical considerations would get in the way of actually using its structure to time-travel.

HOWEVER, THERE ARE GENERALIZATIONS OF MY UNIVERSE THAT INCORPORATE ELECTROMAGNETISM

... AND IN THESE UNIVERSES, CHARGED OBJECTS SHOULD BE ABLE TO TIME-TRAVEL WITHOUT USING SO MUCH ENERGY

E   *B*

The main point, however, is that the laws of physics in our world *do allow* a Gödelian structure – even if it doesn't *actually* have one.

Gödel Against Tenses

Gödel thought that the spacetime he discovered tells us something important about the nature of time; namely, that it doesn't exist! Let me explain. Like McTaggart, Gödel seemed to hold something like the tensed theory of time in mind. The tensed theory, recall, said that the non-relational present moves, turning the unreal future real. Earlier we saw the trouble that special relativity gives this theory of time. For instance, Putnam and others argued that the relativity of simultaneity proved this view of time false. If special relativity caused so much trouble, imagine what Gödel spacetime does.

Therefore, time travel appears incompatible with the tensed theory of time.

Another Problem for the Tensed Theory

Second, and perhaps worse, Gödel spacetime is also impossible to carve up into one series of different instants of time. There is no way of telling the story of Gödel spacetime with a "first instant" marching on to its conclusion in a "final instant". The problem with special relativity was that there were too many ways of doing this. The problem with Gödel spacetime is that there isn't a single one!

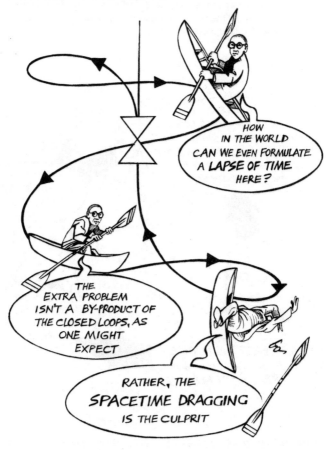

Time travel by itself doesn't affect the carving up of spacetime into spaces at successive times, as cylinder spacetime shows. Cylinder spacetime permits a time-travel loop through every single point, like Gödel spacetime, but it can be neatly sliced into spaces at successive times . . .

OF COURSE, **OUR** WORLD IS **NOT** GÖDEL SPACETIME...

... BUT THAT'S NO COMFORT

... SINCE THE LAWS OF NATURE OF OUR WORLD **ALLOW** GÖDEL SPACETIME...

... THAT IS BAD ENOUGH!

For that means that our world could have been Gödel spacetime had the matter and energy been distributed slightly differently in the beginning. But a "small" difference like that would also mean the difference between having (tensed) time and not having it. How can this be? It can't, Gödel said, so there must not be time in our world.

Was Gödel Wrong?

As with McTaggart's argument, one can read this argument as an attack on the tensed theory of time, not an attack on *time in general*. And some have questioned Gödel's reasoning, too. Why couldn't the existence and nature of time depend on the distribution of matter and energy, they ask? Many important things do depend on it . . .

FOR INSTANCE WHETHER SPACE & TIME ARE **FINITE** OR **INFINITE**

NOT TO MENTION WHETHER WE **EXIST**

But I think Gödel's point was that nothing in physics tells us how matter–energy distributions could give rise to the passage of time, whereas we do know how different matter–energy distributions give rise to space and time being finite or infinite.

Since 1949, it has emerged that there are plenty of time-travel-permitting spacetimes besides Gödel's allowed by the laws of general relativity. Furthermore, physicists have speculated about various methods of actually building or generating paths that go back in time.

One idea by **Frank Tipler** is that of an infinite-sized cylinder . . .

If this works with finite cylinders, and if one day we had the ability to arrange neutron stars – which are very massive and rotate very quickly – we might be able to give this a try. But there are many big "**ifs**" here.

Cosmic String Theory

Another idea is by **J.R. Gott** who showed that an entity known as a "cosmic string" might be able to produce the paths needed for time travel. Cosmic strings are hypothetical relics of the Big Bang – extremely thin filaments of pure energy that stretch the width of the universe.

Wormholes in Spacetime

And yet another idea, vigorously pursued by **Kip Thorne** (b. 1940) and his colleagues in California and by the Russian **Igor Novikov** (b. 1935), is that one may time-travel through "wormholes" in spacetime. Wormholes provide the means of travel used in Carl Sagan's science-fiction novel *Contact* (which was turned into a Hollywood film (1997) starring Jodie Foster). In fact, the work by Thorne's team was apparently inspired by Sagan asking Thorne for a physically possible means of travelling very quickly through space!

The main concept is easy to see. A wormhole is a tunnel, made out of spacetime, between two different points of spacetime. Thinking of spacetime as a rubber sheet again, we can see that a very massive object creates a long "throat" in spacetime. If the closed end of the throat were open and connected to another piece of spacetime, we would have a wormhole. The tunnel would be a shortcut between two different points.

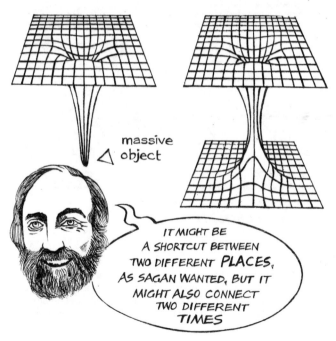

massive
△ object

IT MIGHT BE A SHORTCUT BETWEEN TWO DIFFERENT **PLACES**, AS SAGAN WANTED, BUT IT MIGHT ALSO CONNECT TWO DIFFERENT **TIMES**

Wormholes May Not Allow Travel

The possibility of wormholes has been known since almost the beginning of general relativity. However, because gravity is an *attractive force*, it always wants to close the throat of wormholes.

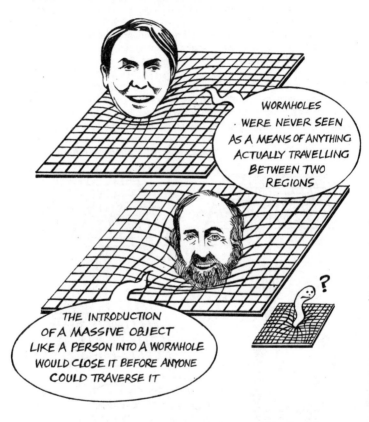

WORMHOLES WERE NEVER SEEN AS A MEANS OF ANYTHING ACTUALLY TRAVELLING BETWEEN TWO REGIONS

THE INTRODUCTION OF A MASSIVE OBJECT LIKE A PERSON INTO A WORMHOLE WOULD CLOSE IT BEFORE ANYONE COULD TRAVERSE IT

The progress Thorne and others are making lies in speculation that would allow one to open a wormhole and then keep it open long enough for someone to pass through. The group known as the Consortium, led by Thorne and Novikov, also does interesting work on the consistency of these scenarios.

We've seen that relativity allows for a variety of methods of time travel. Opposing this work, in a sense, is a theorem by the Cambridge physicist Stephen Hawking. There is of course no real danger of changing history. But Hawking thinks that general relativity, plus some reasonable guesses about how matter and energy are distributed in the world, will prohibit time travel.

Interesting theorems along these lines are difficult and thus rare. So, for now at least, the burden seems to rest more on Hawking and other opponents of time travel to show it is *not* possible than for others to prove that it *is*.

Exotic Possibilities For Time

Time travel is not the only curious temporal property permitted by some general relativistic spacetimes. Let's now look at some other strange possible features of time – first, the idea of "non-orientable" time. Non-orientability is best understood by again thinking of a piece of paper. Cut a strip from a piece of paper and draw little arrows on the paper – with all the arrows pointing in the same direction and the ink dark enough so that you can see the arrows from both sides of the paper.

Now tape together the two ends.

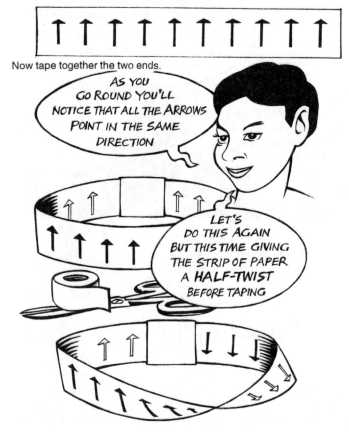

This time you'll notice, if you were a little person walking around the strip of paper, that at a certain point what was once *up* is now *down* . . .

And this happens without any funny business. There are no tears, bumps or stretches of the paper (we're imagining that the paper just came this way, without having to be taped). This piece of paper represents a surface known as a *Möbius strip*, named after the German mathematician and astronomer, **August Ferdinand Möbius** (1790–1868). A Möbius strip is non-orientable, which means that it flips right hands into left hands and up arrows into down arrows.

Möbius Twist in Space

If spacetime were non-orientable *in space*, this would have the
consequence that you could (say) fly from the Earth in a spaceship with a
flag pointing from your ship and – without ever turning over – come back
to Earth upside down with the flag pointing in the opposite direction. Just
imagine the flag is the arrow on the piece of paper.

THE MÖBIUS
TWIST TURNS THE
SPACESHIP OVER WITHOUT
ANYTHING TOUCHING
IT

IF WE
INHABIT SUCH
A SPACETIME, THEN BEING
RIGHT-HANDED OR LEFT-HANDED
WOULD ONLY MAKE SENSE
LOCALLY

Möbius Twist in Time

Spacetime might also be non-orientable *in time*. Think of the arrows on the Möbius strip indicating the direction in which the local future lies (the direction in which, say, acorns grow into trees and human beings age). Then at a certain point in your travels the past and future might exchange places! The Möbius twist in this case is oriented in time.

Again, very strange, but possible. Because the Möbius strip is not intrinsically curved – the paper is not stretched or shrunk in any way – we can see that this feature is possible even in flat spacetimes.

Branching Time

Now let's turn to another idea that time can *branch*. The idea is that space might divide into two (or more) pieces, with time running up each of these separate pieces . . .

We would then have more than one timeline (even putting aside relativistic considerations).

Does Space "Run Out"?

The great French philosopher **René Descartes** (1596–1650) would have thought this impossible. Like some Ancient Greek pre-Socratic philosophers, he didn't think it made sense to ever "run out" of space.

NOWHERE IN YOUR TRAVELS WILL YOU ENCOUNTER AN "EDGE" OF SPACE

WE MUST THEREFORE ASSUME THAT SPACE IS EXTENDED WITHOUT LIMIT

And he took this to imply that space is infinite, since without edges or boundaries, he didn't think that space could be finite. Space's finitude, in turn, suggested that space must be an indivisible unity. For if space is infinite, he thought, there is "no room" for another one not already connected to this one.

Finite Space Without Limit

None of Descartes' inferences hold in general relativity. Because spacetime can be curved, it's possible that it be spherical, like a ball. We therefore know that space can be extended without limit and yet be finite: a basketball is finite yet an ant on its surface would never bump into any walls or boundaries.

It's even possible to have infinite spaces – say, our cylinder spacetime from above – dividing into two infinite planes. We need not conceive of these spaces taking up "room" in a larger embedding space, because, as we saw earlier, we need not think of spacetime as occupying a higher dimensional space.

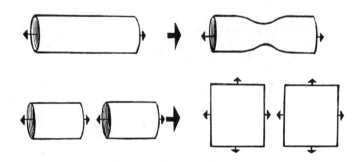

So space need not be a unity, and it's possible that time can branch (though relativity puts severe restrictions on exactly what is allowed).

Geroch's Theorem

Before leaving this topic, we should pause to appreciate an interesting connection among time travel, branching time, and non-orientability. The Chicago physicist **Robert Geroch** (b. 1942) proved in 1967 that if the "topology of space" changes with time – which, for our purposes, happens when space divides and thus when time branches – then the spacetime (if closed and without edges, like the sphere spacetime above) must also either have *paths for time travel* or be *non-orientable* in time.

Eternal Recurrence

Another possibility is that of *eternal recurrence* – the idea that each state of the world recurs an infinite number of times. The idea goes back again to pre-Socratic philosophers, but it was made famous by the German moral philosopher **Friedrich Nietzsche** (1844–1900).

This is almost what happens to Bill Murray's life in the Hollywood movie *Groundhog Day* (1993). About this theory, the comedian Woody Allen remarks that, if it is true, he bitterly regrets having watched the Ice Capades (a US ice show of controversial entertainment value).

Travelling to Big Bang

Is eternal recurrence theory possible according to what we know? The answer is a weak "yes." The English physicists **Roger Penrose** (b. 1931) and Stephen Hawking proved in the 1960s and 1970s that a general relativistic world like ours must have had its matter and energy increasingly concentrated as we go into the past. Indeed, at a certain point, the paths of any traveller would run out. This feature is commonly taken to be evidence of a point – a so-called *singularity* – where matter and energy are so concentrated that the forces would be infinite and the very point ill-defined.

The relevance of the Big Bang is that it might be thought to rule out eternal recurrence. If our universe began roughly 12–15 billion years ago in the Big Bang, then it hasn't been cycling infinitely long. But the singularity theorems do not rule out eternal recurrence.

A Philosophical Objection

Time would be defined from one phase of the universe to another in a kind of "sausage link" universe.

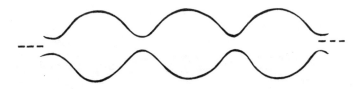

There is of course a simple philosophical objection to eternal recurrence. Consider what Nietzsche says – then consider how this would be *perceived*.

Today some physicists speculate about similar spacetimes, but ones wherein the repeated phases are not identical to one another.

Closed and Open Time

Time, we think, can be either open (linear) or closed (cyclic). It's worth pointing out, however, that in relativity even this distinction is *relative to the observer*. Our useful friend, flat cylinder spacetime, can illustrate this fact. Consider rockets A and B.

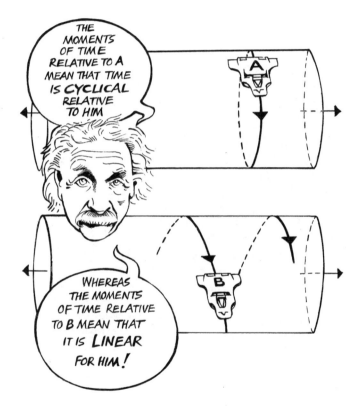

Depending on the observer, the very same spacetime could have an infinitely long time or a finite closed one. Again, it is a question of perception.

To Sum Up, So Far . . .

We've investigated a number of strange properties that time might have and the question of whether the laws of Nature will allow them or not. Surprisingly, general relativity appears to allow all of them – and to throw in a few we didn't anticipate for good measure.

WHETHER ANY OF THESE STRANGE PROPERTIES MANIFEST THEMSELVES WILL DEPEND HEAVILY ON THE EXACT MAKE-UP OF OUR UNIVERSE

BUT IT'S INTERESTING TO KNOW THAT RELATIVITY ALLOWS THEM

We will now turn our attention away from relativity and spacetime to the *material contents* of our universe and its relationship to the direction of time.

The Direction of Time

In Philip K. Dick's novel *Counter-Clock World* (1967), the direction of time flips in 1986, putting the Earth into what its inhabitants call the "Hogarth Phase". Named after the scientist in the story who predicted that "time's arrow" would change direction, the Hogarth Phase is a period in which many processes happen in reverse order.

During this time, the dead call from their graves to be excavated. People clean their lungs by "smoking" stubs that grow into mature cigarettes. White coffee spontaneously separates into black coffee with milk, and so on.

"Irreversible" Processes

Although such reversals of "time's arrow" may occur in works of fiction, they don't seem to happen in the real world. The processes of Nature behave in a temporally asymmetric manner.

Time Reversal Invariance

However, fundamental physics says that these strange reversed processes *could* happen. The fundamental laws of Nature appear to be *time reversal invariant*. This means that the laws of Nature are indifferent to the past and future directions of time. Amazingly, it appears not to be contrary to the fundamental laws of physics that milk spontaneously separate from coffee, or air in a room spontaneously concentrate in a small corner.

Consider the following two pictures . . .

Even before labelling the time order, we know immediately which one came earlier and which one came later.

Seeing in Terms of Particles

Now let's look at that same situation, not with our own eyes but with electron microscopes, and let's focus on a small group of the particles in the china shop.

Here the directional nature of the scene disappears. If we think of the particles as small billiard balls governed by classical Newtonian physics, all we see are some colliding this way and some colliding that way. There is no way to say, from only looking at the pictures, which one was taken first.

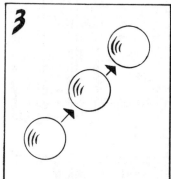 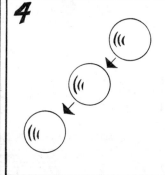

Both orders, 3 then 4 or 4 then 3, are equally acceptable as far as Newton's laws go. But they are also equally acceptable as far as many interpretations of *quantum mechanics* go. Quantum mechanics is the theory that replaced Newtonian physics in the 20th century.

If we took a motion picture of the particles, and then played it backwards, the reversed movie would also display a process allowed by the laws of physics. This is what it means to be time reversal invariant.

Thus, from the point of view of fundamental physics, it's possible for all the bits of broken china to jump up and reassemble themselves as the bull walks backwards out of the shop.

None of this happens, of course, and we might be curious as to why. But we don't yet have an official problem, since lots of things that are possible don't happen. It's possible from the point of view of physics for somebody playing basketball his or her whole life never to miss a shot. But we don't rack our brains worrying why this never occurs. To get a real difficulty, we need more science.

The Science of Heat

In the 18th century, the science of thermodynamics was developed. Thermodynamics is the science of heat. At its inception, it was concerned with the theory behind building ever more efficient steam engines. Thermodynamic processes include the spontaneous transfer of heat from a hotter body to a colder one. If you place a hot object next to a cold object, the hot object will lose some of its heat to the cold object, thereby warming it up. Hot coffee left in room temperature will cool down and slightly warm up its environment.

HUGGING A PERSON COMING OUT OF THE COLD WARMS THEM UP

NOTICE THAT THIS TRANSFER JUST HAPPENS AUTOMATICALLY

You don't actually have to grab heat and move it. "It", whatever it is, moves by itself. And when the temperatures of both objects are equal – when *equilibrium* is obtained – the transfer stops, again spontaneously.

Another common thermodynamic process includes the spontaneous expansion of a gas through its "available volume", the space surrounding the gas through which it can move. If someone came into a room with a flask of poisonous chlorine gas, put the flask in a corner and then opened it, we would know to run away.

Spontaneous Processes

As you can see from these examples, thermodynamics often deals with temporally asymmetric phenomena. Heat in a *closed system* spontaneously goes from hot to cold, never cold to hot. The gas spontaneously expands through its available volume, and never spontaneously contracts. Similarly, in a closed system at room temperature, ice cubes spontaneously melt and puddles never spontaneously turn to ice.

OF COURSE, WE CAN MAKE SOME OF THE REVERSE PROCESSES OCCUR

WE COULD SCOOP UP THE PUDDLE AND PUT IT IN THE FREEZER AND MAKE IT INTO ICE CUBES

But that is cheating, for the refrigerator system is not closed – it draws energy from an outside power source in order to do *work* on the object. Yet the process of ice melting happens spontaneously without any work.

The Law of Entropy

To describe all these asymmetric processes, thermodynamics contains a law – the Second Law of Thermodynamics – based upon work by the French physicist and military engineer, **Sadi Carnot** (1796–1832). The law was phrased several ways, but eventually it was put by the German physicist **Rudolf Clausius** (1822–88) as the claim that the **entropy** of a closed system always increases with time.

THE *ENTROPY* IS A FEATURE OF EVERY BODY

IN THERMODYNAMICS IT IS THE HEAT CHANGE DIVIDED BY THE TEMPERATURE

For us the important point is that the entropy *increases* when the above processes go in the order in which we actually find them and it would *decrease* if they were to go in their reversed order. The Second Law therefore rules out the weird reversed processes. And so that's that for the possibility of these reversals, right?

The Problem of Newtonian Particles, Again

No, the reversals aren't entirely ruled out. The problem is that ice cubes, warm bodies and gases are made of Newtonian particles. Actually, they're constituted of *quantum fields*, but let's neglect this complication. Gases, warm bodies and ice cubes are, let's assume, *nothing more than* Newtonian particles in motion.

And if Newtonian physics declares that the reverse motions are possible, then that has to mean that the reverse motions ruled out by the Second Law are in fact possible. The Second Law cannot be 100 per cent strictly true.

How then do we explain the existence of thermodynamic behaviour in terms of Newtonian particles?

Statistical Mechanics

Enter the great physicists **Lord Kelvin** (1824–1907), **James Clerk Maxwell** (1831–79), **Ludwig Boltzmann** (1844–1906) and **J. Willard Gibbs** (1839–1903), among others. The theory they invented is called *statistical mechanics*. It was vindicated by fluctuations away from thermodynamic values in very tiny systems.

The chief insight underlying statistical mechanics' explanation of the Second Law can be easily illustrated. Imagine we have two boxes, A and B, and 20 billiard balls, numbered 1–20.

THINK OF THE MANY WAYS WE COULD DISTRIBUTE THESE BALLS BETWEEN THE TWO BOXES

FOR EXAMPLE WE COULD PUT ALL 20 IN A AND NONE IN B, OR VICE VERSA

OR WE COULD PUT NUMBERS 1, 7, 13 & 20 IN A AND THE REST IN B

A Statistical Asymmetry

Boltzmann noticed the following interesting asymmetry here. There are many more ways to spread the balls between the two boxes evenly than not evenly. For instance, there is only one way to put all the balls in A and none in B. But there are over 15,000 ways to put five in A and fifteen in B (1–5 in A, the rest in B, 3,4,13,16,18 in A, rest in B, . . .).

And there are more than 180,000 ways to put ten in A and ten in B!

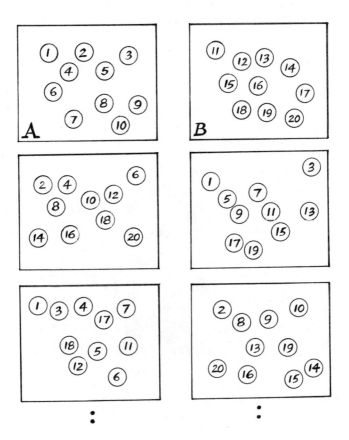

If we thought of each arrangement as equally probable, then it's very, very probable that the billiard balls would be in a 10–10, 9–11 or 11–9 arrangement, and very unlikely (probability = 0.000001) that they would be in a 0–20 or 20–0 arrangement.

Is Reversal Probable?

Now let's return to our gas – it could be contained in either of the two boxes. Newton tells us that the gas could perfectly well remain in one box even when the wall between them is removed.

The correspondence to thermodynamics should now be obvious. The equilibrium distributions, say 10–10, 9–11, 11–9, 8–12, 12–8, are the most likely. The non-equilibrium distributions are unlikely. Newton's reversals are possible, grants Boltzmann, just as a 20–0 distribution is possible – but they are improbable.

In fact, they're monstrously improbable. A typical observable gas has not 20 particles as above, but more than 10^{23} (10 followed by 23 zeroes) particles! The probability of the particles being evenly spread between the two boxes is overwhelmingly high. In this way Boltzmann consistently combines thermodynamic behaviour with Newtonian physics.

Chlorine gas staying in its open flask is extremely improbable, since there are so many other places for it to travel. To stay where it is would require millions of extremely unlikely collisions between its molecules. Similar considerations show that ice melting at room temperature and heat flowing from hot to cold are also most probable.

The Most Probable State of Entropy

Entropy becomes a measure of how probable a state is. Very probable states, like the 10–10 arrangements of billiard balls, have high entropy, whereas the 0–20 arrangements have low entropy.

But with Boltzmann's theory explained, we should now see a problem. Boltzmann's explanation of thermodynamic behaviour basically just uses Newtonian mechanics and some maths. But this explanation of particle behaviour is time reversal invariant. Nothing in what was said by either picks out the direction of most probable behaviour as the direction that we call "the future".

The Loschmidt Paradox

In fact, it seems we *can* run the argument backwards. Given an improbable present state – say, the recently opened flask of chlorine – it follows from the above reasoning that *earlier states* – just like later states – were *more probable states*, too. Given the identification of probable states with high entropy, however, this means that Boltzmann's explanation says that entropy was higher *before* the flask was opened.

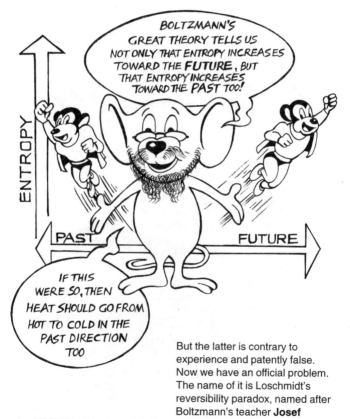

But the latter is contrary to experience and patently false. Now we have an official problem. The name of it is Loschmidt's reversibility paradox, named after Boltzmann's teacher **Josef Loschmidt** (1821–95) who pointed out a similar consequence with one of Boltzmann's earlier attempts to explain time's arrow.

In What Direction Does Entropy Increase?

The Oxford mathematician and physicist Roger Penrose draws the Loschmidt problem like this . . . Boltzmann's statistical mechanics predicts that entropy will increase in *both* directions, whereas experience tells us that it increases in *only one* direction (call it the future direction).

In our game with two boxes and 20 balls, we will eventually get "low entropy" distributions such as 5 in box A and 15 in box B. We only have to wait a long while. When discussing all the constituents of the universe – which number much greater than 20 – we have to wait a great while longer. But given an indefinite length of time, eventually we will expect low entropic fluctuations.

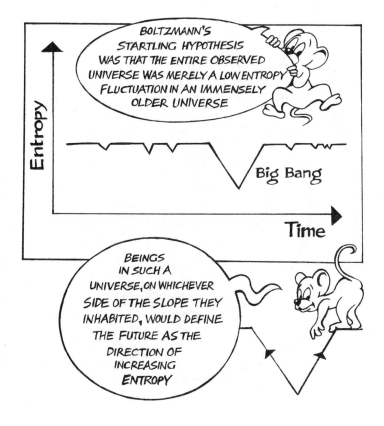

The Universe's Statistical Development

Boltzmann explains why we see the direction of time going in only one direction. Thinking in terms of the two boxes, we know that if we started off in a 5–15 distribution we would next expect to go into a 6–14 distribution, then 7–13, then 8–12, etc. until equilibrium. That is what is happening in our world. The universe is one big game of two boxes. Most of the time it lives in the 9–11 and 10–10 distributions. But then – improbably but expectedly given enough time – it jumps into a 5–15 distribution.

Then Boltzmann's reasoning tells us that entropy ought to increase from this starting point.

The Earth, by contrast, spits out much degraded energy into the solar system.

And we don't expect reversed processes to occur because the state the universe is in right now is already so unlikely. For it to get even more unlikely is, well, very unlikely.

The Boundary Conditions of the Universe

Boltzmann's startling suggestion contains the kernel of what must be the best answer to the above puzzle. That kernel of truth is that the only way to escape the puzzle is to suppose that the *beginning* of the observable universe – and not the *end* of the observable universe – is one of very low entropy. Today no one accepts Boltzmann's full answer. Boltzmann was writing before all the evidence we now have about a Big Bang creating the universe roughly 12–15 billion years ago. Today we think that what Boltzmann considered a small fluctuation from an "older" universe – the observed universe – is in fact all there is.

Penrose estimates it to have the chances of 1 out of $(10^{10})^{123}$! We answer Loschmidt, therefore, by introducing temporally asymmetric boundary conditions.

An Unlikely Hypothesis

Another problem with Boltzmann's answer is philosophical. Compare the likelihood of two hypotheses . . .

H1: *Boltzmann's view that the whole observed universe is a huge and rare fluctuation from normal equilibrium.*

H2: *that the observed universe fluctuated into existence* ten years ago, *complete with all the traces of a longer past (e.g. memories, dinosaur bones, old-looking geological formations, etc.).*

So Boltzmann's answer, carried to its logical extreme, puts us in the awkward position of saying that it's most likely that the universe just popped into existence a moment ago.

Why Does Entropy Really Increase?

Not everyone is happy with the modern version of Boltzmann's theory. One source of discomfort is the monumental improbability of our universe starting off in the state it did. Surely, some say, there must be a deeper explanation of the time arrow than merely the fact that the universe began in a very special state? To this end, physicists and philosophers have proposed numerous ways of escaping this difficulty. But they almost always commit a fallacy that the Australian philosopher **Huw Price** calls a "temporal double standard."

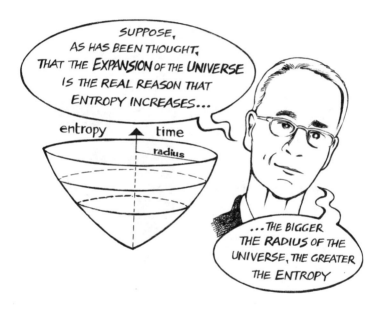

SUPPOSE, AS HAS BEEN THOUGHT, THAT THE **EXPANSION** OF THE **UNIVERSE** IS THE REAL REASON THAT ENTROPY INCREASES...

entropy   time
radius

...THE BIGGER THE **RADIUS** OF THE **UNIVERSE**, THE GREATER THE **ENTROPY**

Suppose also, as we have seen, that the laws of physics are time reversal invariant. The low entropy beginning is then explained by the smallness of the radius at the Big Bang and shortly thereafter. And the increase of entropy is explained by the existence of a small radius constraint at the Big Bang and the *lack* of such a constraint at the *end* of the universe.

The "Temporal Double Standard"

The problem is that, to be consistent, we must treat both ends of the universe the same way. If – as cosmology suggests is possible – we live in a world that ends in a Big Crunch (the temporal opposite of a Big Bang), then we had better say that entropy is low at the other end too. In this case we have posited a world wherein the direction of time flips. Not to do this would be to commit Price's "temporal double standard".

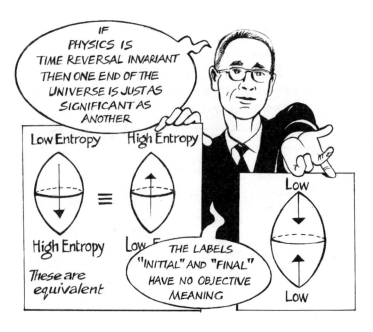

Either we explain the low entropy condition in a way that applies to both ends of the universe, says Price, or we can't explain it at all. We can't say that a constraint operates at the initial end but not the final end if the theory is time reversal invariant.

This is only one example of the double-thinking that plagues the literature on time's arrow. An early and common example of this mistake is the idea that anti-thermodynamic behaviour of the kind found in a time-reversed region of the universe is unlikely. Think, for instance, of the miraculous correlations that would have to occur for the reverse of an egg falling from a table to occur.

Energy from the ground would have to hit the floor in just the right places, with just the right strengths, and just the right directions, to enable all the fragments of egg to bounce up from the floor and reassemble on the table.

Unlikely!

Yes, but the important point to see is that from the reversed temporal perspective – which is equally legitimate according to physics – this sort of unlikely behaviour is occurring all the time all around us.

A Reversal of Time's Arrow

Assuming that in the past entropy was low and in the future it will be high doesn't rule out the possibility that "time's arrow" could flip in certain regions of the universe. After all, it could be that after reaching a point of high entropy in the future, it then turns over and heads for a final state of low entropy again.

If the future endpoint is also of low entropy, then we would expect entropy increase from both ends of the universe towards the middle of the universe . . .

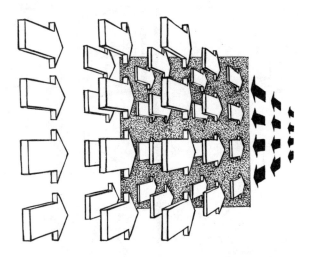

That is, from our perspective, we would expect the direction of time to turn around.

Whether this will happen is not something we know, since we know next to nothing about the nature of the universe's endpoint.

Communication with the Time Reversed?

Many interesting questions arise about such a universe, not the least being whether we could communicate with "reverse" people evolving from what we call the future endpoint. Many philosophers and science-fiction writers have tackled this topic. Consider a race of people living in a galaxy who think what we call the Big Crunch is the Big Bang beginning of their universe.

Suppose our galaxy "meets" this galaxy somewhere in the middle of the entire spacetime.

Could we communicate with them?

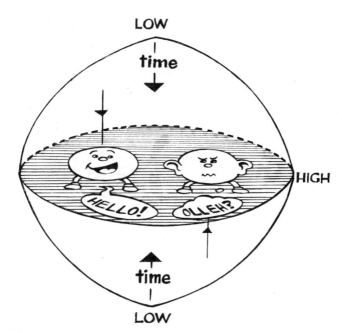

At first you might think "no" because the beginning of our message would be the end of it from their perspective. But there is no reason why they couldn't play the recording backwards, just as people used to play records backwards to find "hidden" messages.

Time-reversed Communication

But there is something odder involved. (Beware, thinking this through may cause a brief headache!) If you send them a question, say at time t^1, asking them "Is the universe finite?", they will receive it at time t^2. From your perspective t^1 is earlier than t^2; but from their "reversed" perspective, remember, t^2 is earlier than t^1. If they answer the question, we may receive it at t^1 or even before.

That is, we may receive the answer before we even ask the question . . .

In much the same way as we encountered when discussing time travel.

The philosopher **Murray MacBeath** has thought of clever ways of avoiding this obstacle to communication involving time-delayed messages. And the science-fiction writer **Greg Egan** based an entire story on this idea . . .

(The headache should be kicking in about now . . .)

Quantum Gravity: The End of Time?

In the past two decades, a new challenge to time – to its very existence – has arisen in the field of quantum gravity. It is widely believed that our best theory of the very large, *general relativity*, conflicts with our best theory of the very small, *quantum field theory*.

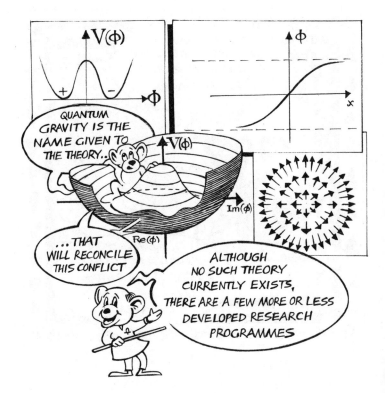

Two major approaches are *superstring theory* and *canonical quantum gravity*. We cannot explain the details here, but we can sketch the problem known as the "problem of time" that plagues the standard, or "canonical", approach.

The Wheeler-DeWitt Equation

The problem of time in canonical quantum gravity is easily explained – there isn't any! The main equation of this theory, the so-called Wheeler-DeWitt equation, arises from applying quantum mechanics to general relativity. Yet this equation lacks any time dependence.

However, many believe that they can either rectify the situation or live with it. Plus, the theory has a number of virtues, so it would be premature to give up on it quickly.

Rounding-up of Positions

It is appropriate that we conclude this book with a discussion of this problem – not only because it details the most recent speculative ideas about time, but also because in the reactions to the *timelessness* of the Wheeler-DeWitt equation, we hear echoes of nearly all the positions on time so far discussed.

Some people (including the author of this book) suggest that we add an "external" time to the equation.

IN THIS VIEW THE WHEELER-DE WITT EQUATION IS AN INCOMPLETE ARTICULATION OF THE PROBLEM

A SECOND FUNDAMENTAL EQUATION IS ADDED

THIS ADDITIONAL EQUATION NATURALLY INTRODUCES A NEW TIME VARIABLE

On its surface (though it can be interpreted differently) this external time is a bit like the **absolute** time of Newton.

The Perfect or "Master" Clock

Others think that the mathematical description or "formalism" of canonical quantum gravity is complete, yet that it conceals an undiscovered time variable or "master clock" within it. These people comb the formalism looking for something that might play the role of time, i.e. some clock with respect to which we can explain the changes we observe.

But in the attempt to find an internal clock in the formalism, we see a view of time a bit like that of either Leibniz or Poincaré.

The Inexistence of Time

Finally, there are those like the English physicist **Julian Barbour** (b. 1937) who think that this theory helps spell the end for time. He believes that the formalism is telling us something deep – namely, that time doesn't exist. This is of course reminiscent of McTaggart's and Gödel's views on time. And as with McTaggart and Gödel, we must ask . . .

Perhaps Barbour, a relationalist and conventionalist about time, can be read as saying not that time doesn't exist, but that time is a skimpier entity than you thought (for instance, not tensed, not Newtonian, not linear and not even fundamental).

A Better-known Mystery

Time is as much a mystery for us as it was for St Augustine. But science and philosophy have sharpened the questions. Thanks to statistical mechanics, we can now formulate the problem of the direction of time. Thanks to general relativity, the science of spacetime, we can now rigorously investigate questions about time travel, branching time, and so on. And thanks to philosophy, we understand the logical geography better: for instance, we know that time might be absolute, relational, conventional, tensed or tenseless, or unreal.

Further Reading

For further reading about clocks, see David Landes' *Revolution in Time* (Viking, 2000). *Longitude* (Fourth Estate, 1996) by Dava Sobel is an entertaining history of the discovery by Harrison of clocks accurate enough to be used by sailors to determine longitude.

For the early philosophical history of time, see Richard Sorabji's *Time, Creation and the Continuum* (London: Duckworth, 1983).

A good start on the philosophical topics of absolutism, conventionalism and relationalism can be found in Hans Reichenbach's *The Philosophy of Space and Time* (NY: Dover, 1958), Lawrence Sklar's *Space, Time and Spacetime* (LA: University of California Press, 1974) and Bas van Fraassen's *An Introduction to the Philosophy of Time and Space* (NY: Columbia University Press, 1985).

For more on time as the fourth dimension, see George Gamow's *One, Two, Three . . . Infinity* (NY: Dover, 1988) and Rudy Rucker's *Geometry, Relativity and the Fourth Dimension* (NY: Dover, 1977).

A completely non-technical but accurate and sophisticated introduction to general relativity is Robert Geroch's *General Relativity from A to B* (University of Chicago, 1981).

Some good essays in the philosophy of time are found in Robin LePoidevin and Murray MacBeath's *The Philosophy of Time* (NY: Oxford University Press, 1993).

On the direction of time and related issues, see Paul Horwich's *Asymmetries in Time* (Cambridge, MA: MIT Press, 1987), Huw Price's *Time's Arrow and Archimedes' Point* (NY: Oxford University Press, 1996), and Steven Savitt's collection, *Time's Arrow Today* (Cambridge: Cambridge University Press, 1995).

Roger Penrose's discussion of the Second Law of Thermodynamics in *The Emperor's New Mind* (Oxford: Oxford University Press, 1989) is recommended.

For time travel, Paul Nahin's book, *Time Machines* (NY: Springer-Verlag, 1999, 2nd ed.), is a must. The book is an entertaining discussion of time travel as it appears in physics, metaphysics and science fiction, and it contains an almost exhaustive bibliography. The books by Horwich and Savitt also contain good discussion of time travel.

And for the implications quantum gravity may have for time, see Julian Barbour's *The End of Time* (London: Phoenix Paperbacks, 1999) and Craig Callender and Nick Huggett's collection, *Physics Meets Philosophy at the Planck Scale* (NY: Cambridge University Press, 2001).

Good Internet resources include the entry on Time in the Internet Encyclopedia of Philosophy (www.utm.edu/research/iep/t/time.htm) and the entry on time travel and modern physics in the Stanford Encyclopedia of Philosophy (plato.stanford.edu/entries/time-travel-phys).

The Authors

Craig Callender received his PhD in Philosophy in 1997 from Rutgers University with a thesis entitled *Explaining Time's Arrow*. He lectured at the London School of Economics from 1996–2000 in the Department of Philosophy, Logic and Scientific Method, rising to the position of Senior Lecturer. He is now Assistant Professor of Philosophy at the University of California, San Diego. He has published in philosophy, physics and law journals and specializes in the philosophical foundations of physics.

Ralph Edney trained as a mathematician, and has worked as a teacher, journalist and political cartoonist. He is the author of two graphic novels, and the illustrator of *Introducing Philosophy* and *Introducing Fractal Geometry*. He is also a cricket fanatic.

Acknowledgements

Craig Callender is very grateful to Lisa Callender and Pat McGovern for their generous help with a draft of this book.

Index